Second Edition

GIS
Fundamentals

Second Edition

GIS
Fundamentals

Stephen Wise

CRC Press
Taylor & Francis Group
Boca Raton London New York

CRC Press is an imprint of the
Taylor & Francis Group, an **informa** business

CRC Press
Taylor & Francis Group
6000 Broken Sound Parkway NW, Suite 300
Boca Raton, FL 33487-2742

Library of Congress Cataloging-in-Publication Data

Wise, Stephen.
    GIS fundamentals / Stephen Wise. -- Second edition.
        pages cm
    Includes bibliographical references and index.
    ISBN 978-1-4398-8695-3 (pbk.)
    1. Geographic information systems. 2. Geospatial data. I. Title.

G70.212.W58 2013
910.285--dc23   2013023893

Visit the Taylor & Francis Web site at
http://www.taylorandfrancis.com

and the CRC Press Web site at
http://www.crcpress.com

Contents

Preface

When I set out to design a GIS course for third-year undergraduates in the Department of Geography at Sheffield University, I decided to include some material on how GIS worked at a fundamental level. The students were already familiar with how to use GIS from earlier courses, and so the next obvious step was to understand how they worked. An analogy which I often use is that they knew how to drive the car – now it was time to learn how the internal combustion engine works. The problem I faced was that there was very little reading that I could provide for them which was at a level that they could understand. Some of the early work in GIS was undertaken by geographers, but modern GIS systems draw on ideas developed in mainstream computer science. This makes the literature difficult to follow for anyone without a background in this discipline, although the techniques used to store and process spatial data are not fundamentally difficult to understand. I decided to solve my problem by writing my own book, and this is how the first edition of *GIS Basics* came to be produced. In the years since it was published, there has been an explosion in the use of spatial data so that, from being a specialist area, GIS technology is now something which is used via technologies such as the Internet, GPS and mobile phones by huge numbers of people on a daily basis. It therefore seemed timely to update the book to reflect some of these changes. The name has been changed to *GIS Fundamentals* to better reflect the fact that this is not an introductory text. However, the aim remains the same, which is to provide an introduction to some of the main ideas and methods that underpin GIS in simple, everyday language.

Learning Features

The book does not attempt to be comprehensive or exhaustive and is deliberately written to be readable and accessible. When a detailed explanation would interrupt the flow of the text, it is provided in a separate box. The book can be used by students in two ways:

1. To deepen your knowledge of GIS by understanding in more depth how GIS works. If this is your aim, then you may find that the book is sufficient in itself. While not exhaustive, the coverage of material is quite broad.

2. To help you understand the GIS literature better. As well as explaining the techniques and methods for handling spatial data, the book also tries to introduce some of the language and the approach used by computer scientists and contains a separate glossary of terms. This will all help you to tackle some of the more advanced literature.

Suggestions for further reading are given at the end of each chapter, with full bibliographic details at the end of the book. There are not many references to Internet sites, but this is not because there are not any useful ones. The problem is that Internet sites come and go, and so the only ones mentioned are those that seem likely to have some permanence. However, an Internet search on any of the topics covered in this book will usually bring up some useful material. In particular, the coverage of computer science topics on Wikipedia is excellent. Articles are accurate and well written and often contain examples to help illustrate the ideas.

Structure of the Book

The book is really best used by being read in a serial manner because later chapters do build on ideas from earlier chapters. At a first reading, this would be a difficult book to dip into at random. The chapters can be split into the following sections:

Chapters 1 and 2: These provide some general introductory material on how computers work, the nature of programming languages and databases. They also introduce the particular issues caused by the 2D nature of spatial data.

Chapters 3–11: These cover the methods used to handle different types of spatial data in the following order:

Chapters 3–5: Vector data

Chapters 7–8: Raster data

Chapters 9–10: Surface data

Chapter 11: Network data

The alert reader will have noticed that Chapter 6 is missing from this list. This is because it covers the subject of how computer science measures the efficiency of algorithms. This is something that is of key importance but which is hard to understand without some prior knowledge of how programs to handle spatial data are written. Therefore, this material comes after the coverage of vector algorithms rather than in the introductory chapters.

Chapters 12 and 13: The final two chapters make a return to two important general topics which are relevant to all spatial data types. A key issue in all data processing is how to provide rapid and flexible access to large sets of data. The two-dimensional nature of spatial data poses some particular problems in this respect, and Chapter 12 is devoted to the problems and a range of possible solutions. Chapters 3 through 11 describe a series of GIS tasks for which there are efficient solutions that are guaranteed to give the optimum results. However, there are tasks in which it is not possible to find the best possible solution in any reasonable time. Such problems call for the use of special methods called heuristics that aim to find solutions that are good enough rather than being optimal, and these are described in Chapter 13.

Changes from the First Edition

As with anything to do with technology, the change in the field of information technology, and especially in the use of spatial data, has been enormous. When the first edition was published in 2002, spatial data could only be handled by specialist GIS software, and doing anything with it required a lot of training in using this software. Now, spatial data handling is an everyday operation for many people. People are navigating using satnavs on their car, which make use of GPS positioning and route planning software. Anyone with access to the Internet can browse maps of the Earth's surface that make use of enormous databases of maps, imagery and information about places. And these services can increasingly be accessed via mobile phones that possess more computing power than was used to put men on the moon. It therefore seemed timely to update the book to reflect some of the changes that have occurred.

Perhaps the most important changes have been the requirement to provide rapid access to very large volumes of data, which means that traditional database technology is now more important in GIS than it used to be. The coverage of databases in the first edition was quite brief, but it has now been expanded to form a chapter in its own right. The chapters on vector and raster have also been updated to cover developments in handling these data types in standard database systems. The chapter on indexing spatial data (Chapter 12), which was quite short in the first edition, has been completely rewritten and now describes the main methods used to improve the efficiency of accessing data in modern GIS systems. As a result of this, there is now more coverage of a variety of tree data structures that are important in indexing information. Another change that is partly a result of the greater use of spatial data on the Internet is a greater use of imagery in GIS, and so the chapter on raster data (Chapter 7) has been expanded to include this.

Route planning is now a widely used technique, and so the discussion of this in the chapter on networks (Chapter 11) has been expanded to cover methods designed for planning routes across large networks very rapidly. As a result, the material on the travelling salesman problem has been moved to a new chapter (Chapter 13) that addresses problems for which no algorithm exists, which can provide a result in a reasonable time and where heuristic methods must be used.

And finally, the new name! Changing the name of the book was not a decision that was taken lightly. It was really prompted by comments that the original name *GIS Basics* suggested an introductory GIS text, whereas the book was clearly aimed at those with quite a lot of GIS knowledge. The original name had its origin in a series of articles which covered a small selection of topics and which went under the name 'Back to Basics'. The name was intended to suggest that the articles were describing some underlying principles, but it was also a joke on a somewhat discredited policy of John Major's Tory government in the United Kingdom. Hopefully, the new name *GIS Fundamentals* captures the aim of the book equally well but without giving the suggestion that this is a book for beginners.

Acknowledgements

Thanks to Irma Britton at CRC Press who first suggested a second edition of this book and who then helped through the writing process and to various staff at CRC Press for help in the production process. The quality of the book was improved enormously by extremely helpful suggestions from Peter van Oosterom and Nick Tate and from Marc van Kreveld and Steve Maddock who helped me understand the Bentley–Ottmann algorithm.

The maps of Chatsworth maze in Chapter 1 are reproduced with the kind permission of the Chatsworth House Trust. Many of the figures in the book are derived from Ordnance Survey digital map data, as indicated by the copyright beneath them, and are reproduced with kind permission of Ordnance Survey under licence NC/01/576. The ASTER L1B data used in Chapter 9 were obtained through the online Data Pool at the NASA Land Processes Distributed Active Archive Center (LP DAAC), USGS/Earth Resources Observation and Science (EROS) Center, Sioux Falls, South Dakota (https://lpdaac.usgs.gov/get_data). The LiDAR data in the same chapter were downloaded from the Channel and Coastal Observatory site www.channelcoast.org. The yEd graph editor that was used to draw the diagrams of tree data structures is a free graph drawing package produced by yWorks and available from http://www.yworks.com/en/index.html. Many of the other vector figures were drawn using an open-source vector editor called Inkscape, which can be downloaded from http://inkscape.org/.

Author

Stephen Wise earned a degree in geography in 1976 at Bristol University and then spent four years as a postgraduate student in London. Between 1980 and 1990, he worked as a computer programmer in University Computer Centres in London and Bath before moving to his current position at Sheffield University, where he is a senior lecturer in the Geography Department. His teaching and research are mostly concerned with GIS (geographic information system), although he also teaches physical geography. In his spare time he is a keen musician and a member of the Down Trodden String Band. His wife and two daughters do not know the slightest thing about GIS.

1

Introduction

If you are familiar with using a GIS (geographic information systems) package such as ArcMap or MapInfo, you will be aware that one of the things you must learn is how to 're-think' a problem into a form which can be tackled by using the GIS software. For example, you may be interested in whether traffic has an effect on the vegetation near roads. This is a spatial problem; however, to use GIS and to study it, we will have to be more specific about what sort of effect we might expect, and what we mean by 'near'. We might decide to concentrate on species diversity using results from a series of sample plots at varying distances from a road. A very simple analysis would be to compare the results for plots within 10 m away from a road with those which are further away than this. In GIS terms, we would create a 10-m buffer around the road and produce summary statistics for sites within and for those outside it.

In the same way as learning to use a GIS involves learning to 'think' in GIS terms, learning about how GIS programs work at a more fundamental level involves learning to 'think' like a computer. To understand this, it is very useful to know a little about how computers actually work. Most people will be familiar with thinking about computers in terms of how to use the operating system to run programs, how to create and save information in files and how to organise those files into folders. Most people will also have an understanding of the main physical elements of the computer such as the screen, the keyboard, external devices such as printers and connections to the outside world via a network. However, to understand the material covered in this book, you will need some understanding of how a computer works at a more fundamental level than this – it is rather like the difference between knowing how to drive and understanding how the internal combustion engine works.

This first chapter will provide some background information on how computers work at the most fundamental level. The chapter begins with an informal description of solving a spatial problem designed to give you an intuitive understanding of how computers work. This is followed by a more detailed description of how data are represented on the computer and how these data can be manipulated using a programming language.

1.1 How Computers Solve Problems

The problem we will consider is that of finding a route through a maze, such as the one shown in Figure 1.1 (which is actually the maze at Chatsworth House in Derbyshire). This is quite a simple maze, and most people will be able to trace a route to the centre fairly easily.

Part of the reason that this is a simple task is that we can see the whole plan of the maze laid out before us. Now imagine that you are standing in the gardens of Chatsworth House at the entrance to the maze. All you can see is the opening in the hedge. When you enter, you will be at a junction with a path leading to the left and a path to the right. It is clear that if you are going to find your way to the centre, you will have to adopt some form of a strategy. With many mazes, a method which works is to place your hand on one of the walls, and walk along keeping your hand on the same wall all the time. If you place your left hand on the left-hand wall in the Chatsworth maze, then the route you take will be as shown in Figure 1.2.

This strategy has worked – we have found a route to the centre. What is more, to make it work we only have to process one piece of information at a time – does the wall on the left carry straight on or turn a corner? This is important because computers operate in this one-step at a time manner.

The actual work in a computer is carried out by the central processing unit or the CPU for short. Given the complexity of modern computers, it is perhaps surprising to learn that at the most fundamental level the CPU is only capable of doing a few things, such as adding two numbers together. If we want to add together three numbers, we have to add the first two, save the result and then add this to the third one. So how can a computer perform

FIGURE 1.1
The maze at Chatsworth House. (Copyright Devonshire Collection Chatsworth.)

FIGURE 1.2
Route through the maze using the left-hand rule. (Copyright Devonshire Collection, Chatsworth.)

complex calculations? By breaking them down into a sequence of simpler calculations. And how can it solve complex problems? By breaking them down into a sequence of simpler steps. The resulting sequence of steps is called an algorithm.

Our maze-solving strategy is a good example of an algorithm. It is designed to solve a problem, and does so with a series of steps. What is more important to note is that the steps are unambiguous – there is no subjective judgement to be used, simply a set of rules to be blindly followed. The maze example is also useful to illustrate some other important ideas in algorithm design.

The first is that this simple algorithm will not always work. It is very easy to design a maze which cannot be solved by this algorithm, and an example is shown in Figure 1.3.

Here, our simple algorithm will simply revolve around the outer wall and return to the entrance. We will need a more sophisticated strategy. For instance, we will need to be able to detect that we have returned to the same place, to avoid circling around the outer wall for ever. This raises another issue – to solve a problem on the computer, we need some way of storing the relevant information in the first place. There are two stages to this. The first is to create a representation of the information which will tell us what we want to know and which we can store on a computer. In the case of the maze, the obvious thing would be to store the plan of the walls, as a set of lines, and this could certainly be done. However, what is actually important is the paths and particularly where they join one another. A diagram of the paths and where they join for Figure 1.3 is shown in Figure 1.4.

Both Figures 1.3 and 1.4 are examples of what are sometimes loosely referred to as 'Data Models' (a more thorough description of data modelling is presented in the next section). These are still pictorial representations

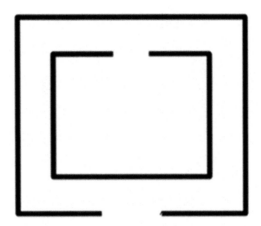

FIGURE 1.3
A maze which cannot be solved using the simple algorithm.

though and so the next stage would be to work out a method for storing them in a computer file, which requires what is sometimes called (again rather loosely) a data structure. It is the data structure which the computer will use to solve our problem, using whatever algorithm we eventually come up with. For instance, a better algorithm than our wall-following approach would be to use the path data model. Using this, we would need an algorithm which traces a route from the origin point to the destination. At each junction, we would need to decide which path to try first. We would also need to keep track of which paths have been tried, to avoid trying them again, so the data

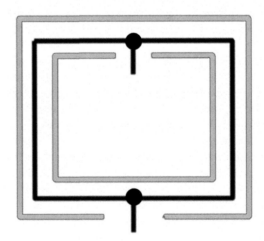

FIGURE 1.4
Diagrammatic plan of the paths and connections between paths in Figure 1.3. The maze itself is shown in grey.

structure will need to be able to store an extra piece of information for each path, labelled 'Has this been tried or not?' It can be seen from this that the design of data structures and algorithms are closely linked.

The key developments which have led to the creation of modern GIS software were the design of data structures and algorithms for handling spatial data. The maze example has shown that algorithms are sequences of steps designed to solve a problem, where the information for the problem is stored in a suitable data structure. There is one other important feature of algorithm design, which is nicely illustrated by the original solution to the Chatsworth House maze (Figure 1.1). This algorithm worked, but it was very inefficient since it involved walking up and back down several of the paths. With a small maze this would not be too important, but with a much larger one, it would clearly start to matter. From this, we can see that algorithms should not simply work, but work efficiently.

This first section has provided an informal introduction to the main themes which will be covered in the book – the design of data structures and algorithms for handling spatial data. Before these can be covered, however, we need to consider the important topic of data modelling in a little more detail.

1.2 How Computers Represent the World: Data Modelling

We have seen that in order to process information about the real world on the computer, we must have a representation of that information in a form which the computer can use. The real world is infinitely complex, so we will first need to select the information which is relevant to the problem at hand and then decide how to organise this to be stored on the computer. This process of going from some real-world phenomenon to a computer representation of it is termed 'data modelling' and it can be considered as a series of steps as shown in Figure 1.5 which is based on Worboys and Duckham (2004).

We have already seen examples of the first two stages in this process with the example of the maze problem. The original map of the maze was an example of an Application Domain Model – a representation of the system of interest in a form comprehensible to people who are familiar with the real-world system. It is a simplified model of the real maze and like all models it represents the most important elements, in this case the walls and paths, leaving out unimportant details, like the height of the hedges or the type of surface on the paths.

For the maze-solving problem, this model could be simplified even further since all that was needed was the connections between the paths and these could be represented as shown in Figure 1.4. This is an example of what mathematicians call a graph, and is an important way of representing spatial data as we shall see later. In effect, this is a model of one element

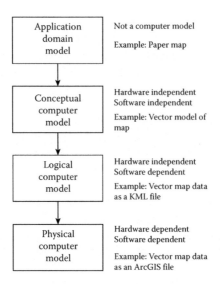

FIGURE 1.5
Data modelling steps. (After Worboys, M.F. and Duckham, M. 2004. *GIS: A Computing Perspective* (2nd edition). Boca Raton, FL: CRC Press.)

of the original map, and forms the conceptual computational model in this example. It is a more abstract representation than the original map, but it is not in a form which can be directly stored on a computer.

The next two stages in the process will be described in a moment, but first a word about these first two stages. To be useful for handling a wide range of spatial data, GIS must be able to represent a wide range of application domain models and conceptual computational models. Several earlier works on GIS were addressed at representing application domain models which were two-dimensional and static. The paper map is the classic example since maps of all kinds are models of the real world which take certain elements of the natural or built environment and provide a simplified representation of them. Many of the key developments in GIS have involved the design of the conceptual computational models and logical computational models for representing spatial data derived from maps on the computer. Another major area of development was the ability to use data from remote sensing, which represents a rather different application domain model, but which also presents a static, 2D view of the world. Many current developments in GIS are concerned with extending GIS software to be able to deal with those things which are difficult to represent on a map or remote sensing image such as change over time, variation in the third dimension, or non-map-based views of the world such as the view of a city from street level. However, the map-based view of the world is still a useful one, and is widely used in spatial data applications, and this will form the focus of the material covered in this book.

To return to our description of data modelling, the application domain model for GIS can be considered to be the map, of which there are two main types. The topographic map is a general purpose map showing a range of features of the Earth's surface – examples are the Ordnance Survey Landranger and Pathfinder maps in the United Kingdom, or the sort of maps produced for tourists showing places of interest, roads, towns and so on. The other type of map is the thematic map, of which a soil map is a good example. This is not a general map of the area, but a map of the variation or pattern of one thing or 'theme'. Another good example are maps derived from census information showing how the characteristics of the population vary across an area.

In both cases, two types of information are being shown on the map. First, the map shows us where things are – the location of settlements, woodlands, areas of high population density, for example. Second, it shows us some of the characteristics of those things – the name of the settlement, whether the woodland is deciduous or coniferous, the actual number of people per hectare. The location is shown by the position of the features on the map, which usually has some kind of grid superimposed so that we can relate these positions on the map to real-world locations (e.g., measured in latitude and longitude). The characteristics of the features are usually indicated by the map symbolism, a term which includes the colours used for lines or areas, the symbols used for points and filling areas and whether the line is dashed or dotted. Hence, the settlement name would be printed as text, the woodland type would usually be indicated by using a standard set of symbols for each woodland type, and the population density would be shown by a type of shading with the actual values indicated on a key.

To store this spatial data, the computer must be able to hold both types of information – where things are and what they are like, or to use the correct terms the locational and attribute data. This requires an appropriate conceptual computational model, and in fact two are in common use – the vector and raster models. Much of what follows will be familiar to most readers, but it forms a necessary prelude to the description of the logical computational models which follow. To illustrate the principles of the vector and raster models, let us take the example of the topographic map shown in Figure 1.6, and see how the information shown on it would be stored in each case.

In a raster GIS, an imaginary grid is laid over the map. Each cell in the grid, called a pixel, is examined to see what feature falls within it. As computers work best with numbers, a code is used to represent the different features on the map – 1 for coniferous woodland, 2 for deciduous woodland and so on. Therefore, all the pixels which cover areas of coniferous woodland will have a number 1 stored in them. Figure 1.7 shows this process for part of the map – on the left a grid is superimposed on the map, on the right is the grid of numbers. Blank cells in the grid would actually have a zero stored in them, but these are not shown here.

The end result of this will be a grid of numbers which can easily be stored in a computer file. The grid can be used to redraw the map – each pixel is

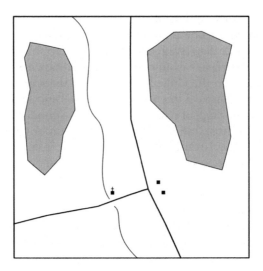

FIGURE 1.6
Imaginary topographic map.

coloured according to the code value stored in it – dark green for conifers, light green for deciduous woodland and so on – producing a computer version of the original map. We can also do simple calculations using the grid – by counting the number of pixels containing deciduous woodland we can estimate how much of the area is covered by this type of landcover.

In assigning the codes to the pixels, there will be cases when one pixel covers more than one feature on the original map. This will occur along the

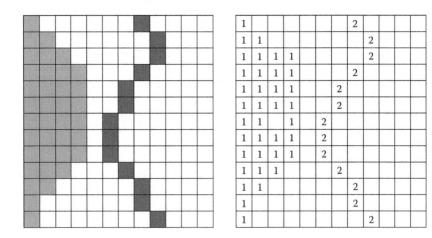

FIGURE 1.7
Raster version of part of the map in Figure 1.6. Left – diagrammatic view of the raster map. Right – the grid of numbers which actually represents the map in a raster system.

boundaries between land-use classes, for example. In these cases, whichever item occupies the majority of the pixel will have its value stored there. Another example is when a road and a river cross – we cannot record both in the same pixel. In this case, rather than try and store all the data in a single grid, we use separate grids for different types of features, so that in this case roads and rivers would each be stored in their own grid. Note that this means that we can use the GIS to look for places where roads and rivers cross, by laying one grid on top of the other.

The end result of this process is a grid or a series of grids of numbers which represent the features on the map – one grid is often referred to as a layer. The numbers in the pixels therefore represent the attribute data but how is the location of the features stored in this kind of system? Since the pixels are all square, as long as you know the location of one point in the grid, and the size of the pixels, the location of any other pixel can be worked out fairly simply. Most raster systems therefore store this information for each raster layer, sometimes in a special file which is kept separately from the layer itself.

In a vector GIS, each object which is to be stored is classified as either a point, a line, or an area and given a unique identifier. The locational data from the map are then stored by recording the positions of each object, as shown in Figure 1.8 which shows the vector locational data for the map in Figure 1.6.

The buildings have been stored as points and the location of each one is recorded as a single pair of coordinate values. It is clear that we have only stored part of the information from the map – for instance, two of these points are houses, while the third is a church. This attribute information is

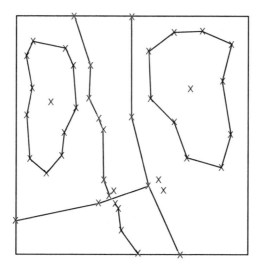

FIGURE 1.8
Vector version of the topographic map in Figure 1.6.

normally stored separately, in a database, and linked to the locational data using the identifier for each point. Lines, such as the roads and rivers, need to have multiple coordinates representing points along the line, except in special cases where it may be possible to use a mathematical function to represent the course of the line. Areas, such as the two woodlands, are represented using a line to represent the boundary, and a point marking some point inside the area. This is not essential, but is useful for locating labels when the area is drawn.

Vector and raster, which are often loosely referred to as data models, could more appropriately be regarded as conceptual computational models using the terminology shown in Figure 1.5. This means they each provide a model of a map in a form which can be stored on the computer, but without being tied to any particular software package or type of computer. This is the purpose of the next stage in the modelling process – the design of a logical computational model. As a simple example of what is meant by this term, consider the example of a raster layer, which consists of a set of values in regular grid representing values of some phenomena across space. Almost all computer languages have a data structure called an array, which can be used to hold two-dimensional matrices of values. The syntax used in the C programming language for example is as follows:

```
int landuse[2][2];
```

This sets up an array of two rows and two columns which can hold integer values. The C array is therefore one example of a logical computational model – it is a representation of the raster model in a form specific to one piece of software (the C programming language). Once the C program is compiled, then we reach the final stage of the modelling process – actual files on a particular type of computer, or the physical computational model. Although these are important, they are the concern of software designers, and need not concern us here.

In the Worboys scheme, the term logical computational model is used to refer to the way in which the vector and raster models can be represented in software. The term data structure is also commonly used in the GIS literature for this same stage in the modelling process, especially when discussed in conjunction with the design of algorithms to process the data and this usage will be adopted for the rest of this book.

1.3 The Structure of a Computer

It is now time to consider the internal workings of a computer in a little more detail. Fortunately, it is possible to do this in a way which does not involve understanding the actual electronics inside a computer. Despite the wide variety of computer makes and models, from small portable computers to large supercomputers, their internal structure can be represented as five main elements as shown in Figure 1.9.

At the heart is the CPU which is the part of the machine which can actually perform computation and this is where we will start. As stated earlier, despite the complex range of functions a computer can perform, the CPU is actually only capable of a few operations. For example, it can take two numbers and perform basic arithmetic such as addition or division on them or it can compare two numbers and determine which is the larger. So how can this be enough to perform complex operations? One way of understanding this is to imagine using a hand-held calculator to perform a calculation. The example which will be used is the calculation of the trajectory of a bullet fired from a gun. If air resistance is ignored, the trajectory of the bullet is a mathematical shape known as a parabola and it is possible to derive the equation for the distance the bullet will travel before hitting the ground

$$R = \frac{v^2 \cdot \sin(2\theta)}{g} \tag{1.1}$$

where
θ = Angle of the gun relative to the ground
v = Speed of the bullet as it leaves the muzzle
g = Acceleration due to gravity

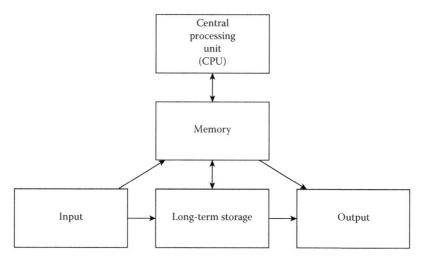

FIGURE 1.9
Schematic diagram showing the main elements of a computer.

Incidentally, this is not a randomly chosen example. It was the need to make this sort of calculation for the artillery which led Charles Babbage to conceive of the idea of designing a machine to do the calculations and so led to the development of the Difference Engine, which is usually considered to be the first computing machine.

To calculate R given an initial velocity and an angle using a calculator, we would have to break the calculation down into a series of steps:

1. Calculate 2 times the angle theta.
2. Find the sine of this value.
3. Put the result in memory – call it M.
4. Calculate the square of the velocity v.
5. Retrieve M form memory and multiply by v squared.
6. Divide the result by the acceleration due to gravity 9.81.

This is quite a lot of steps for what seems a simple calculation. Now imagine that your calculator does not have anything except the basic arithmetic operations and a memory. Step 4 will still be simple – instead of using the square function you just multiply V by itself. But what about step 2? How do you calculate a sine without a SIN button? The answer is that you do what used to be done before computers – you use a formula such as the Taylor series

$$\sin(x) = x - \frac{x^3}{3!} + \frac{x^5}{5!} - \frac{x^7}{7!}\cdots \qquad (1.2)$$

Calculating a single sine has now become even more work than the calculation of the ballistic range. Each term in the series involves multiplying alpha by itself numerous times, multiplying integers together to produce the factorial and then dividing these two numbers together. However, the important point is that it can be done.

This example illustrates the fact that complex calculations can be performed using a machine which can only do basic arithmetic as long as we can break the calculation down into a series of small steps. One of the pioneers of modern computers, the English mathematician Alan Turing, introduced the concept of what he called an 'a-machine'. He imagined a machine which was fed with a strip of tape marked out into separate boxes, each containing a single symbol. At any given moment, there would be just one symbol inside the machine with the remainder of the tape stretching out on either side. The machine is capable of doing just two things, namely to read and write the symbols and to move the scanning head to the left or right along the tape. It is programmed by having a table which instructs the machine on what each symbol means it should do.

For example, a symbol A might be code for 'Erase the current symbol and move two places to the left'.

Unbelievable as it may sound, this machine is capable of computation given the correct combinations of symbols and instructions. What is perhaps even more unbelievable is that any computer program, no matter how complex, could in theory be implemented using such a machine, which is now generally referred to as a Turing machine. The Turing machine was never intended as an actual device for computation, but as a means of thinking about what can and cannot be computed by a machine. We will return to this issue of computability at the end of the book.

The second point which the ballistic calculation example and the Turing machine both illustrate is that if we need to break operations down into very small steps, we always need some way of storing parts of the calculation while we work on the rest. You will see that in Figure 1.9 there are two boxes below the CPU labelled memory and long-term storage. Memory is where we can store information while the computer is turned on – it is like the memory in our calculator in that when we turn the machine off it 'forgets' everything that is stored. This is fine for doing calculations but we also need some way of storing information in some more permanent way which is why we also have some form of long-term storage such as a disk drive. Finally, we have some way of inputting information and some way of getting output.

Although the diagram of the computer in Figure 1.9 is very simple, it is generally applicable to all general-purpose computers from the very earliest to the current generation of laptops and personal computers. Something else which is a fundamental feature of computers is the relative speed at which information is passed between the different elements of the system. The processing of information within the CPU is very fast. CPU speeds are measured in terms of the number of operations which it can perform per second. For example, this book is being written on a 3-GHz machine. This means that it has what is called a clock cycle of 3 billion times a second (where a billion is 10^9) – 3 billion times a second, something happens inside the CPU. To actually do something useful, like adding two numbers, may take a few clock cycles, but this is still very fast indeed. By contrast, moving information from memory to the CPU and then back to memory is slower than this, partly because the distance it must travel is further and partly because it takes longer to find the correct information in memory. Moving information from long-term storage is slower still for the same reasons. Slowest of all is the input and the output. On a single machine, this involves mechanical devices such as a mouse, keyboard, or printer. When the computer is connected to a network, input may mean fetching information from a completely different computer, which is what happens every time you access an online mapping service or book a train ticket online.

We have established that we have a hierarchy of speeds from the fast CPU to the slow transfer of information across networks. There is also an opposite hierarchy of information content. The CPU is fast, but can only handle a small number of pieces of information. Memory can handle more, but not as much as external storage or the potentially infinite storage of the Internet. These ideas are fundamental to the way in which computers are designed and programmed. To be useful, computers must work quickly; throughout this book, we will see the numerous ways in which this is achieved. A useful analogy to bear in mind when considering this is to imagine yourself reading this book in a well-stocked university library, but one without Internet access, and trying to answer a question such as 'Who is generally credited with formalizing the details of the architecture of modern, digital computers?' If you happen to know this, then you will answer immediately – John Von Neumann. This is because the information is in your memory. If you do not know it, you will look in the library first, which is the equivalent of looking in local, long-term storage. How quickly you find the answer depends on two things – whether there is a book which contains the information and how well the library collection, and each book is indexed. If you cannot find the answer, you will need to expand your search to other sources of information – visiting other libraries, or nowadays, doing an Internet search, both of which will take more time. In principle, then the speed with which you can answer the question depends on how well organised your sources of information are, and how quickly you can access them, and these two principles govern how quickly computers can operate.

Let us now turn to consider how the information on which the computer operates is stored. In our description of the CPU, it was said that one of the things a CPU could do was to add two numbers together. How are these numbers stored? Most people know that computers use the binary system. This means that all numbers are stored using a base 2 numbering system. In decimal arithmetic, we use the digits 0–9. Once we reach 9, if we add 1, we write the result as 10. This is a way of writing numbers in which the position of each digit has a meaning. The '1' in the number 10 is actually shorthand for '1 lot of 10'. A number such as 254 is therefore a way of recording that we have '2 lots of 100', '5 lots of 10' and '4 lots of 1'. The binary system works in exactly the same way except that we only have two digits – 0 and 1. Thus, the binary number 111 is '1 lot of 4', '1 lot of 2' and '1 lot of 1' – in other words 7 in decimal arithmetic.

The CPU has a series of what are called registers, each of which can store a single number in binary format. In a 16-bit computer, such as those which used the Intel x86 processors, each registers can store numbers with 16 binary digit, or bits. Two of the registers are called AX and BX. If we wanted to add 1 and 2, we would store each number in a register:

```
AX:  0000000000000001
BX:  0000000000000010
```

Remember that '2' is '10' in binary. The CPU could then add these, saving the answer in register AX, giving

```
AX: 0000000000000011
BX: 0000000000000010
```

But how do we tell the computer that we want it to add these two values together? The answer is that each of the operations the CPU can perform such as addition or subtraction is given a numerical code called an opcode. For instance, on the x86 processors code, 00010101 (which is 21 in decimal) is one of the codes for addition. The computer will also need to know which registers to use and where to store the result and so there will be codes for this information too. And how is all this stored? In another 16-bit 'number'. The term 'number' is potentially misleading here since we are storing quite a complex set of information but doing so using the same number of bits that we use for numbers. This is because all information on the computer is stored in a fixed number of bits. On most computers, there are two main units of storage – the 8-bit byte and a longer unit called a word which on older computers was 16 bits but on more recent models is 32, 64, or even 128 bits. So, our computer instructions are stored in 16-bit words.

So, we now know that the CPU can take information, stored in words, and do operations such as addition on it. We also know that the instructions for those operations are also stored in words. And, all these – the instructions and the data – are stored in memory ready to be passed to and from the CPU. In order to be stored when the computer is turned off the same information is saved to long-term storage and then read back into memory when it is needed.

We have seen that the computer works by breaking a problem down into very tiny steps which operate on one or two pieces of data at a time. However, it would be very tedious if it was necessary to write programs by describing every single tiny step and also very difficult to explain how programs work by going into this level of detail. It would be like describing a recipe for making a cake by describing the ingredients and the changes that happened during cooking in terms of the chemical changes happening at the molecular level. Fortunately, in the same way that cooking can be described in much simpler terms, computers can be programmed in a way which is far easier to understand. Throughout the book, algorithms will be described using what is called pseudocode which is quite similar to some of the languages which are actually used to program computers.

1.4 Pseudocode and Computer Programming

To give a sense of the difference between the low-level operations described in the previous section and how most programs are written, let us start with the simple addition of 1 and 2. We saw how these numbers would have to be converted into binary, copied into registers, added together and then copied back to memory. In a typical programming language, this same addition could be performed using something such as the following:

```
A = 1 + 2
```

The letter A is what is called a variable. It can be used to store a number, a date, or a letter. The language provides a syntax which is similar to mathematics to add 1 and 2 and store the result in variable A. We could also have written

```
A = 1.0 + 2.0
```

This appears to be the same, but in fact numbers with decimal points are stored in a completely different way from whole numbers, as we shall see later. However, when writing the program, we do not need to worry about that – this will all be handled by the special program which takes our code and turns it into the detailed sequence of instructions which will actually run on the computer.

The statements above would be valid in some languages, but not others. For instance, in C, a semicolon is needed at the end of the line. In fact, a C program to add 1 and 2 would look like this

```
int main(void)
{
int a;
a = 1 + 2;
}
```

The line int a; tells the computer that variable a will be used to store integer numbers. The next line does the addition and all the other lines are simply required by the conventions of the language.

However, all these details get in the way of trying to explain how an algorithm works, and so for this purpose it is usual to use what is called pseudocode rather than a real programming language. Pseudocode is something which often resembles a programming language but which is not bound by strict rules of what can and cannot be done. The only rules are that

- The pseudocode should convey the operation of the algorithm being described clearly. This is often best done by using a syntax which is similar to a programming language, but it is also quite normal to use a written description if this makes things clear.
- It should be possible for a programmer to implement the algorithm in any suitable language. This means the pseudocode should not rely on features which are only available in one language.

The easiest way to introduce the pseudocode which will be used in this book is to show how it can be used to describe a simple algorithm. The example which has been chosen is an algorithm for finding prime numbers. These are numbers which can only be divided by themselves and by 1. For instance, 4 is not prime, because it can be divided by 2, but 3 is prime. One itself is a special case and is considered not to be a prime number. A simple method of finding primes is called the 'Sieve of Eratosthenes' after the early Greek scientist who first described it. We know that any number which is divisible by 2 cannot be a prime number. Therefore, if we start at 2, every second number is not a prime and can be deleted from consideration. Starting from 3, every third number cannot be a prime and can also be deleted. We already know that 4 is not prime, because it was deleted when we considered multiples of 2. We also know that multiples of 4 will already have been dealt with in the same way. We therefore skip to the next number which has not been deleted from our list, which is 5. In theory, this method would identify all the primes up to infinity. In practice, we can only do the calculations up to a defined limit. If we decide to find all primes up to 100, we only need to apply our method starting with numbers up to 10, the square root of 100. There is no need to consider numbers which are multiples of 11. 11 multiplied by more than 11 will be greater than 100, because both numbers are bigger than the square root of 100. 11 multiplied by anything less than 11 has already been dealt with.

Let us start by writing out this algorithm as a series of steps:

```
1. Make a list of numbers from 1 to 100.
2. Mark number 1 since it is not a prime.
3. Set k = 1
4. Until k equals or exceeds 10 do steps 5 to 7
5. Find the next unmarked number in the list after k. Set k to
this value.
6. Starting from k*2, mark every kth number in the list.
7. Add one to k.
```

Note that this description of the algorithm is effectively already in pseudocode because it contains enough information for a programmer to write a computer program to run it. However, let us convert it into a form of pseudocode which more closely resembles an actual programming language. The format used is based on the one used by Cormen et al. (1990).

First, we are going to need a set of 'boxes', numbered from 1 to 100 representing the numbers. In programming this is called an array, and we will represent it as follows:

```
NUMBERS [1..100]
```

Each box will have to be able to hold a label indicating whether it has been marked or not. The simplest method is to put a NO in every box at the start, and label those which have been marked with a YES. To put a value in one box, we use this syntax:

```
NUMBERS [1] = NO
```

To set all the elements of NUMBERS to NO, we can use the idea of a loop, which repeats an action a set number of times:

```
for i = 1 to 100
NUMBERS[i] = NO
```

Here, i is a variable which can hold any single item of information – a number or a character string. In this case, it is initially set to 1, and then successively to 2, 3 and so on up to 100. Each time, it is used to reference the next element of the NUMBERS array.

Since this is a pseudocode, we probably do not need this detail, so we might equally well say:

```
Set NUMBERS[1..100] to NO
```

since any programmer will know how to do this. Number 1 is a special case, so we mark it YES:

```
NUMBERS[1] = YES
```

We use another variable k and set it to 1.

```
k = 1
```

The next step is to find the first number after k which is not marked. Again, we can use a loop for this:

```
until NUMBERS[k] == NO
k = k + 1
```

The second line of this sequence increases the current value of k by 1. This is repeated until the kth element of NUMBERS is found to be NO. The == sign in the first of these lines is the test for equality, and should not be confused with the single equals sign. Thus, whereas

```
k = 1
```

sets variable k to 1

```
k == 1
```

tests whether the current value is 1 or not, but does not change it. The >= symbol indicates 'greater than or equal to' and != means 'not equal to'.

k is now set to the next prime number. This has to be left unmarked in the list – but k*2, k*3 and so on should all be marked. So, we need to step through the remaining numbers from k*2 marking every kth one:

```
for i = k*2 to 100 step k
NUMBERS[i] = YES
```

Here i is set to numbers between k and 100 in steps of k. When k has the value 2, i will take on values 2, 4, 6 and so on. Each time it is used to set the relevant element in the NUMBERS array to YES to indicate that it is divisible.

Finally, we need to go back and look for the next candidate prime number starting at k + 1. We can now produce a first version of the complete program

```
1.  Array NUMBERS[1..100]
2.  Set NUMBERS[1..100] to NO
3.  NUMBERS[1] = YES
4.  k = 1
5.  until k >= 10
6.  until NUMBERS[k] == NO
7.     k = k + 1
8.  for i = k*2 to 100 step k
9.          NUMBERS[i] = YES
10. k = k + 1
```

There are a number of things we can do to make this easier to follow. Lines 6 and 7 find the next prime number by going through the list of numbers starting at the current value of k and finding the first unmarked one. The algorithm will become much clearer if we replace those two lines as shown below

```
5.  until k >= 10
6.  m = find_next_prime(k)
7.  for i = m*2 to 100 step m
8.      NUMBERS[i] = YES
9.  k = m + 1
10. procedure find_next_prime(n)
11. until NUMBERS[n] == NO
12.      n = n + 1
13. return n
```

The two lines have been replaced by a reference to something called find_next_prime which is defined in lines 10–13. You will see that lines 11 and 12 are the same as our old lines 6 and 7. Now, however, they are contained in what is called a procedure which is called from line 6 in the revised program.

This change has made the main program a little easier to follow because line 6 now simply says that we need to get the next prime without giving the details of how to do this. These details are contained in the procedure which reads in the current value of k and stores it in a local variable called n. It then looks through the NUMBERS array starting at element n until it finds a prime and it returns this value via a return statement. In the main program, on line 6, the variable m gets set to whatever value the procedure returns. Procedures are used a lot in real programming and one of the reasons is that they can clarify a large program by breaking it down into a series of smaller tasks.

At the moment, our pseudocode is designed specifically to find the prime numbers between 1 and 100. It is not difficult to make it more general purpose by letting the upper limit be represented by a variable N instead of being a specific number. This means that we will need to have some way of calculating the square root of N. As we are dealing with pseudocode, we can simply have a statement which sets up a variable called rootN and sets it to the square root of N:

```
rootN = square_root (N)
```

This is self-explanatory – square_root is a procedure which takes a value and returns its square root. For the purposes of illustrating our sieve algorithm, there is no need to write out the details of how the square root of N would be calculated since this is irrelevant to the logic of how we find primes. Finding the square root of a number is something which will often be needed and so most programming languages will provide a standard procedure to do this. This is the second reason that procedures are commonly used in programming languages because, as well as, making programs easier to understand they provide a way of performing commonly needed tasks.

We can now look at a revised version of our whole algorithm:

```
1./* Program to identify prime numbers
2./* using the Sieve of Eratosthenes
3. Array NUMBERS[1..N]
4. rootN = square_root (N)
5. Set NUMBERS[1..N] to NO
6. NUMBERS[1] = YES
7. k = 1
8. until k >= rootN
9.      m = find_next_prime (NUMBERS, k)
10.      for j = m to N step m
11.           NUMBERS[j] = YES
12.      k = m + 1
13./* Print out primes
14. for i = 1 to N
15.      if NUMBERS[i] == NO print i
16./* Procedure to find next prime in the list
17. procedure find_next_prime (NUMBERS[], n)
18. until NUMBERS[n] == NO
19.      n = n + 1
20. return n
```

The lines starting with /* are comments, meant to clarify the program. Even with pseudocode, it is sometimes helpful to include comments on what particular sections of the algorithm are trying to do. A section has also been added to write the results out although since this is not a real program this is not strictly necessary.

This simple program has covered many of the main features of programming languages:

- Storage of information in variables
- Storage of lists of information in arrays
- Idea of looping to repeat an action, with tests to indicate when to end the loop
- Use of procedures to break a task down into smaller pieces

It should provide sufficient background for you to understand the pseudocode examples used in the rest of the book.

Further Reading

Those who are interested in the history of computing might like to look at Chapter 6 of David Evans' book Introduction to Computing (Evans 2011) which is available free online at www.computingbook.org. Anyone who is interested to know how the electronics inside a computer works and how binary numbers are stored and manipulated should look at the free book by Roger Young called How Computers Work (Young 2009) at http://www.fastchip.net/howcomputerswork/p1.html. In 1990, the American National Centre for Geographic Information and Analysis published the Core Curriculum for GIS. This was a series of lecture notes, with associated reading and exercises which included, among other things, lectures on data structures and algorithms. Two versions were produced (NCGIA 2000) and both can be accessed at http://www.ncgia.ucsb.edu/pubs/core.html. Two GIS textbooks (Worboys and Duckham 2004, Jones 1997) have been produced by computer scientists, and therefore contain somewhat more technical material than some of the other texts currently available. Both contain details of vector and raster data structures, and of key algorithms. Worboys and Duckham's (2004) study is generally the more technical of the two, and has a considerable amount on databases, while Jones (1997) has a focus on cartographic issues, including generalisation. The Spatial Analysis Online web site http://www.spatialanalysisonline.com/ gives access to a textbook by De Smith, Goodchild and Longley which covers a wide range of topics, including many of those covered in this volume. The book by Burrough and

McDonnell (1998) also contains some details of data structures. The volume edited by van Kreveld et al. (1997), while not a textbook, contains a series of good review papers covering different aspects of GIS data structures and algorithms. Dale's (2005) study is a good introduction to basic mathematics as used in GIS and will be useful for anyone who requires additional help with any of the mathematical ideas covered in this book. Several introductory texts exist on the subject of computational geometry, which deals with computer algorithms to solve geometrical problems, such as solving mazes. The most accessible for a non-technical audience, and one with a strong emphasis on GIS applications of computational geometry, is de Berg et al. (1997). Godfried Tousaint maintains an excellent guide to computational geometry resources on the web at http://www-cgrl.cs.mcgill.ca/~godfried/teaching/cg-web.html. The Free Online Dictionary of Computing (http://foldoc.org/) is also useful. The general issue of data modelling and the importance of good algorithm design in GIS are covered by Smith et al. (1987). Fisher's short paper (1997) raises some interesting issues concerning what is actually represented in a pixel in a raster GIS, and some of these points are expanded upon in guest editorial in the *International Journal of GIS* by Wise (2000a). The sieve of Eratosthenes algorithm is taken from the description on the following web page: http://www.math.utah.edu/~alfeld/Eratosthenes.html. The web page itself has an animated version of the algorithm in action.

2

Databases

We will start our review of how spatial data are stored and processed by considering the database technology which is used to store large volumes of data by almost all companies and businesses around the world and see whether this can be used. Many people reading this book may well use databases in their daily work and be very familiar with them. However, many more, and the great majority of the public, will know very little about databases and yet almost everybody will have used a database to access information without ever being aware of the fact. If you have ever used a computer system to book a ticket, look up timetable information, book a holiday, hire a car or purchase anything online, then the system you used to search for the item you wanted and book or purchase it will have been a database system. As far as the user is concerned, a database allows you to search through a large amount of information to find the item or product which suits your needs – a ticket for a particular play or a holiday cottage in the right part of the country at the right time of year. For this to be possible, the database designer has had to organise a large and complex mass of information into a structured format and produce a user interface which allows the user to define their needs in simple terms and which presents the items that may be of interest in a clear manner. The fact that most people use these systems without being aware of how complicated they really are is exactly why databases are so powerful and so useful. To fully understand why databases are so widely used, it is worth knowing a little of the history of their development and this is where this chapter will start. Given their widespread use, they are an obvious choice for the storage of spatial data. However, it turns out that this is not a simple matter. To understand why this is the case it is necessary to understand something about the commonest type of database system, the relational database. With this background knowledge, we will then be able to consider the issues which arise when GIS developers first tried to store spatial data in relational databases.

2.1 What Are Databases and Why Are They Important?

As we saw in Chapter 1, the part of the computer which does the actual computation is called the CPU and this is only capable of quite a small number of basic operations, such as adding two numbers together. We also saw that by combining these small operations, it is possible to do the kind of complex tasks of which computers are capable. Each operation is identified by an opcode, indicating what is to be done (e.g., fetch a number from memory) and some additional information, such as where in the memory the number is stored. A series of instructions in this form is called machine code, and it is specific to each type of computer. In the very earliest days of computing, there were essentially no tools to help produce machine code – to program a computer, you simply had to learn the format of the machine code. To give a flavour of how complicated this was, Figure 2.1 shows a representation of the machine code needed to add together the numbers stored in two registers on the Intel x86 family of processors.

Figure 2.1 shows the contents of two 8-bit bytes. Each byte is surrounded by a darker border and each is subdivided into three parts. The first byte starts with six zeroes, which indicates that this is an ADD operation. We will leave the next 1 bit for the moment. The last 1 bit says that we are using 32-bit registers as opposed to 8-bit registers. The whole first byte therefore contains the number 00000011 in binary or 3 in decimal. This is the opcode and this particular code instructs the computer to perform an addition. The second 8-bit byte tells the computer which items are to be added and is made up of three elements:

- 11 is a code which says that the items to be added are both held in registers as opposed to memory.
- 000 is the code for the EAX register.
- 011 is the code for the EBX register.

This will add EBX to EAX and store the result in EAX. The computer knows that we store the result in the first of the two registers because of the other 1 bit in the opcode (the one we skipped earlier). If this had been a zero, the answer would have been stored in EBX instead.

This is how much detail is needed for just one instruction. To produce the correct version of the opcode, the programmer must know whether the computer has 32-bit, 16-bit or 8-bit registers. This instruction only makes sense as part of a program, which would need further instructions to read the numbers in from somewhere into memory and move them to the appropriate registers and save the answer somewhere.

The first development to simplify programming was the introduction of assembly languages, which are a way of making it easier to write machine code. As an example, below is a section of assembly language for the Intel

000000	1	1	11	000	011

FIGURE 2.1
Intel x86 machine code to add together two numbers.

x86 processors which will add 1 and 2 together – the assembly code which corresponds to the instruction above is in bold.

```
; Very simple example of the syntax of x86 assembly code
.data
a  dd 1
b  dd 2
c  dd ?
.code
mov eax, [a]
mov ebc, [b]
add eax,ebx
mov [c],eax
```

The basic instructions now have names instead of just opcodes and there is a simple syntax for specifying the information to be processed. This program starts by putting 1 and 2 into locations in memory which are labelled a and b and setting up an extra location called c for the answer. The a and b values are moved into registers and then add eax,ebx. will perform the actual addition. Finally, the sum is moved into the memory location c.

This will still look very complicated and incomprehensible to most people; but, for a programmer, it is a big improvement on writing machine code directly. However, programmers are still having to operate at the level at which the CPU operates, breaking each task down into a series of very small operations. You also have to understand the way the CPU is constructed at a very detailed level. This makes programming very difficult and restricts the number of people who are able to use a computer. In addition, assembly languages are still specific to the exact CPU being used – change computer and you might have to learn a new assembly language.

A large improvement came with what are called third-generation languages (machine code and assembly languages being the first two generations). As an example, here is the equation for calculating the range of a ballistic, which was introduced in Chapter 1.

$$R = \frac{v^2 \cdot \sin(2\theta)}{g} \tag{2.1}$$

And here is how this would be calculated using a third-generation language such as C

```
r= (pow(v0,2)*sin(2*alpha))/g
```

Compared with assembly language, this now looks much more comprehensible. You have to know that * means 'multiply', pow(v0,2) means 'raise v0 to the power of 2' and '/' means divide – but at least the whole thing looks more like an equation. These languages were possible because of the development of compilers, which are special programs that would translate the program into machine code, very often by first producing assembly language, which was then further translated into machine code. As well as saving the programmer from having to work at such a low level, it also made it possible for languages to become independent of particular computers. A program written in a third-generation language will run on any computer which has a compiler for that language.

Third-generation languages revolutionised the use of computers and made it possible for many more people to start using them. They are still widely used today and new languages continue to be developed. However, third-generation languages can be quite difficult to use and require a programmer to learn a lot of what might be called general housekeeping to make a program run. To illustrate this, and to show why the next generation of languages developed, here is a C program in full, which will calculate the range of a ballistic and print out the answer.

```
 1. #include <stdio.h>
 2. #include <math.h>
 3. int main()
 4. {
 5. float r,v0,alpha,g;
 6. v0 = 20;
 7. alpha = 2;
 8. g = 9.81;
 9. r = (pow(v0,2)*sin(2*alpha))/g;
10. printf('R =%f\n',r);
11. return(0);
12. }
```

The formula is now on line 9. The eight lines which precede it are all needed to allow the program to correctly calculate the value of r. Lines 6–8 set the values for the input parameters. However, before this can be done, we must tell the compiler that we will be using variables called r, v0, alpha and g and that they will all contain floating point numbers. Line 10 prints out the result. The equation on line 9 makes use of what are called functions, such as sin. For the compiler to be able to find these, we need to include lines 1 and 2. And finally, we have a line to declare that this is the main program (line 3), one to identify the end of the program (line 11) and various brackets and semicolons.

This all seems a lot of work to do one calculation but the power of third-generation languages is that with a little modification this program could be made to calculate r for a wide range of different velocities and angles and print the results as a neat table.

However, it still takes quite a long time to learn to write programs and you still have to learn complicated instructions for producing output or for reading data from files. To illustrate how the next generation of programming languages changed this, we are going to change to a different example. So far, all the languages we have looked at have been general-purpose ones, which could be used to get the computer to do anything it was capable of. One price for this generality is that the languages are complex. In contrast, the fourth-generation languages are designed to be used for one purpose, and they are designed to be easy for non-programmers to use. This is where modern database software comes in. Databases are programs written with the sole aim of managing large amounts of data. They therefore have a well-defined list of things they need to be able to do – input data, edit it, print it out and make selections based on different criteria. In addition, they should be able to allow more than one person to access the information at the same time and should prevent two people trying to change the same information at once. Finally, they should be reliable and have some means of controlling which users can update the information. A database is therefore a very complicated program to write. However, many of the tasks they are used for are actually quite simple – print all sales for the latest month, provide a list of books in alphabetic order and the like. It therefore makes sense to provide a simple means of accessing the data and this led to the development of database languages. The best known of these is structured query language or SQL, which is used for relational databases.

To illustrate, we will return to the idea of booking tickets. The table in Figure 2.2 shows the plays which are on at a particular theatre – you have to imagine that this is just a small section from a very long list covering the whole year. What we want is to be able to search this list to find plays which might interest us – for illustration, we will assume we want to see what Shakespeare productions are coming up.

In a relational database, this information would be stored in a table called PLAYS with each column given the names shown in Figure 2.2. To list all the Shakespeare productions using the database language SQL, we might something like

```
SELECT *
FROM PLAYS
WHERE AUTHOR = 'William Shakespeare'
```

Play	Author	Director	Date	Time
Twelfth Night	William Shakespeare	Trevor Nunn	May 1	7.30
Twelfth Night	William Shakespeare	Trevor Nunn	May 2	7.00
Hedda Gabler	Henrik Ibsen	Jonathan Miller	May 5	7.30
Hedda Gabler	Henrik Ibsen	Jonathan Miller	May 6	7.30

FIGURE 2.2
Simple table of productions stored in a table called PLAYS.

This will produce a neatly formatted list of the two performances of *Twelfth Night*. The SELECT * statement means that all columns in the table will be listed – if we just want the name of the play and the date, we would use

```
SELECT PLAY, DATE
FROM PLAYS
WHERE AUTHOR = 'William Shakespeare'
```

And if we wanted the list in date order, we would use

```
SELECT PLAY, DATE
FROM PLAYS
WHERE AUTHOR = 'William Shakespeare'
ORDER BY DATE
```

Because SQL statements have a simple and predictable structure, it is possible to provide a simple interface for people to use, which means that most users do not even have to learn SQL. For instance, a menu could be provided which lets people search for plays by author name – all you need is some way of letting the user type in the author name and then the program plugs this into a preprepared SQL statement and runs the query:

```
SELECT PLAY, DATE
FROM PLAYS
WHERE AUTHOR = 'Name of Author'
ORDER BY DATE
```

It is important to stress that fourth-generation languages such as SQL have not replaced the earlier types of language, but simply provided a simpler way of using the computer for some tasks. The software that actually reads the SQL statements and processes the data will be written by programmers in a third-generation language such as C++, which will be compiled into assembly language and eventually machine code to run on the computer. Sometimes, when a portion of a package needs to run as quickly as possible, programmers will program directly in assembly language because it is then possible to exploit features of the way a particular CPU has been made.

The key point is that database systems now make it relatively easy for organisations to store and use large amounts of data. They present a relatively simple interface, in the form of SQL, for those designing the database and tools to make even simpler interfaces, in the form of menus, for the general user. When considering how best to store and process spatial data, databases are thus an obvious solution. To understand the issues, which arise in trying to do this, it is necessary to understand something of how relational databases work.

2.2 Relational Database

The relational database stores information in tables such as the one shown in Figure 2.2. A table is an extremely common and simple way to store information. What makes the relational database so powerful is that way the information can be combined from more than one table to answer queries.

If we look at our table of plays, we can see that some information is repeated. For instance, every row for a performance of *Twelfth Night* is going to have the same author and director. What is more, there might be additional information about the play which we would like to include, such as reviews or a synopsis of the plot. Some performances may have special characteristics, such as the presence of a signer for those with hearing difficulties. To illustrate how a relational system provides for these, let us consider the issue of the repeated author and director. In a relational system, the table would be separated into two tables as shown in Figure 2.3.

The separation into two tables saves having to repeat the same information, but the SQL query will now need to access both these tables to list the plays by Williams Shakespeare and the dates of the performances. This is done as follows:

```
1. SELECT PLAYS.PLAY, PERFORMANCES.DATE
2. FROM PLAYS, PERFORMANCES
3. WHERE PLAYS.AUTHOR = 'William Shakespeare'
4. AND PLAYS.PLAY = PERFORMANCES.PLAY
```

Some of this is similar to the original query. Line 3 is exactly the same as before. Line 2 now identifies two tables instead of one and the SELECT statement and line 3 now has to make it clear which table each of the columns

PERFORMANCES

PLAY	DATE	TIME
Twelfth Night	Saturday May 1	7.30
Twelfth Night	Sunday May 2	7.00
Hedda Gabler	Wednesday May 5	7.30
Hedda Gabler	Thursday May 6	7.30

PLAYS

PLAY	AUTHOR	DIRECTOR
Twelfth Night	William Shakespeare	Trevor Nunn
Hedda Gabler	Henrik Ibsen	Jonathan Miller

FIGURE 2.3
Information about productions represented by two tables which have the name of the play as a common attribute.

is in. The key to the query is line 4 because this is where the two tables are joined together. In effect, each row in the PERFORMANCES table has a row from the PLAYS table added to it and the row which is selected is the one with the same value for the PLAY column.

This operation of linking two tables together based on information they hold in common is called a relational join and it is the key to the power and flexibility of relational databases. It means that the way the information is stored is very simple since everything is in tables. It means that information which is the same for many entries can be placed in a separate table. This makes the database smaller; but, more importantly, it means the information only has to be entered once. This reduces the chance of errors and makes it easy to change the information if needed. Finally, the relational join means that other information can be added to the database later. For instance, once a play has opened, comments from reviews can be added in a separate table like the one in Figure 2.4.

To list all the reviews for plays produced by Trevor Nunn, we can run this query:

```
1. SELECT PLAYS.DIRECTOR, REVIEWS.SOURCE, REVIEWS.COMMENT
2. FROM REVIEWS,PLAYS
3. WHERE PLAYS.DIRECTOR = 'Trevor Nunn'
4. AND PLAYS.PLAY = REVIEWS.NAME_OF_PLAY
```

Note that in this case the columns being used to do the join are not called the same thing. This is fine as long as the columns contain the same information.

To allow tables to be joined, it is important that each cell in both tables contains only one item of information. For instance, if *Twelfth Night* was a coproduction between Trevor Nunn and Andrew Lloyd Webber, then a table such as the one in Figure 2.5 would not be allowed.

The director column for *Twelfth Night* now contains a list rather than a single name so this part of the SQL

```
1. WHERE PLAYS.DIRECTOR = 'Trevor Nunn'
```

would not work. The table would have to be stored as shown in Figure 2.6.

Now, each cell only contains one item of information. Note that even though they all contain two or more words, those words make up the name of a single item and this is allowed. When cells only contain one item, they

REVIEWS

NAME_OF_PLAY	SOURCE	COMMENT
Twelfth Night	The Times	'Brilliant'
Twelfth Night	The Guardian	'A clever interpretation'

FIGURE 2.4
Table of reviews which can be linked to either table in Figure 2.3 through the name of the play.

PLAYS

PLAY	AUTHOR	DIRECTOR
Twelfth Night	William Shakespeare	Trevor Nunn, Andrew Lloyd Webber
Hedda Gabler	Henrik Ibsen	Jonathan Miller

FIGURE 2.5
A table which could not be stored in a relational database because of the list of directors' names for *Twelfth Night*.

are said to be atomic and when all cells in a table are atomic, the table is said to be in first normal form.

One of the rules governing relational databases is that all tables must be in first normal form, which has important consequences for storing spatial data as we shall see. There are other characteristics of relational databases which also have important consequences. Relational databases are based on an area of mathematics called set theory and each row in a table is regarded as a member of a set. Hence, in the above example, PERFORMANCES is a set with four members and PLAYS is a set with two. In everyday language, these would be called tables, which consist of rows – however, in the special terminology of relational databases, PERFORMANCES is a relation which contains four tuples. Since each row or tuple contains three items, it is actually a 3-tuple. For clarity though, the words table and row will be used in this book.

Regarding the rows as members of a set has some interesting consequences. First, it means which each row must be different from all the others since if two rows had exactly the same entries for all columns there would be no way to distinguish them. Most databases ensure that tables adhere to this rule by having a column which is used simply to identify the rows often using numbers from 1 to N. The second consequence of regarding the rows as set members is that they have no inherent sequence. So, for example, in the PLAYS table above, *Twelfth Night* is not the first row – this is just the way it has been printed in this book. As far as the database is concerned, *Twelfth Night* is just one of the entries in the table. Finally, the same applies to columns in that these have no inherent order with each being identified simply by its name.

PLAYS

PLAY	AUTHOR	DIRECTOR
Twelfth Night	William Shakespeare	Trevor Nunn
Hedda Gabler	Henrik Ibsen	Jonathan Miller
Twelfth Night	William Shakespeare	Andrew Lloyd Webber

FIGURE 2.6
The table from Figure 2.5 in normalised form.

2.3 Storing Spatial Data in a Relational Database

Let us start by briefly reviewing the characteristics of spatial data. It is assumed that the reader is familiar with this, so this summary will be very brief. Figure 2.7 shows an imaginary topographic map containing points (e.g., the houses), lines (e.g., the roads and rivers) and areas (e.g., the woodlands and fields).

There are two ways to store this information in a GIS – in a vector format or a raster format. Vector views the world as being populated by objects, which are either points, lines or areas. Each object has a location and one or more attributes. In everyday life, we can describe locations in numerous ways, including place names (e.g., London) or by reference to other known places (e.g., near to Trafalgar Square). GIS can only make use of location recorded as a pair of numerical coordinates as shown in Figure 2.7. Here, the location of the church is recorded using the distance in kilometers from the origin of the map. In most GIS applications, location is recorded with reference to a standard coordinate system. In many countries, this will be based on the global UTM system while in others, such as the United Kingdom and the United States, specially designed systems are used. With the advent of global mapping provided via the web, position is often recorded in degrees of latitude and longitude. Point features only require one coordinate pair to record location but lines and areas need a series of points to describe the position of the line or the area boundary. In addition to location, each object will also have one or more attributes.

In a raster layer, the same information is stored by choosing which attribute we are interested in and estimating a value for that attribute in a regular lattice of pixels covering the area. As the pixels are of uniform size, shape

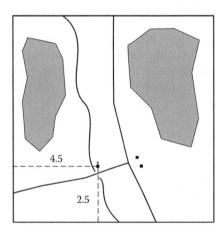

FIGURE 2.7
Imaginary topographic map showing point, line and area features.

ID	X	Y	FC	MATERIAL	NAME
1	4.5	2.5	Church	Stone	St.Saviour's
2	5.8	2.9	House	Brick	1, The Green
3	6.0	2.2	House	Stone	The Larches

FIGURE 2.8
Data for buildings on Figure 2.7.

and orientation, a few pieces of information serve to define the location of any pixel. For example, the coordinates of two diagonally opposite corners of the raster grid and the size of the pixels completely define the grid. A raster representation of the buildings on the map would therefore consist of a grid in which just three pixels would have a value representing 'building' and all the others would contain zero or a value representing an empty cell.

Now, let us see how we might store this data using a relational database. We will start with a vector point layer for the three buildings, which could be stored as shown in Figure 2.8.

There is nothing wrong with this as a relational table. We have ensured that each row is unique by adding an ID. There is no sequence in the rows since each simply describes a separate building. Each row has the same number of columns and each cell is atomic. Finally, there is no sequence in the columns.

However, when we turn to lines, things get more complicated. Figure 2.9 shows data for the two roads on the map. Each road has two attributes – an indication of quality and a measure of the traffic volume. The location has to be stored as a series of points. When joined by straight lines, these will provide an approximation of the road's course. However, this list of *XY* pairs gives us a problem. We cannot store them in a single cell as shown in Figure 2.9 because of the rule that all cells must be atomic.

If we try and solve this by placing each value in a separate cell, we now have a problem because the columns in a relational table have no inherent order. Simply labelling them X1, Y1, Y2, Y2 is no help as a relational database has no way of making use of this information and will see these as unrelated attributes of each object (Figure 2.10).

The only way to store the data is to normalise it as shown in Figure 2.11. Here, the data have been split into two tables. ROADS records the attributes for each road while ROADPTS stores the locational data. Each point is labelled with the road it belongs to and its position in the sequence of points

NAME	QUALITY	TRAFF	XY
A23	Fair	1000	4.5,10.0,4.5,5.7,5.5,,2.5, 6.5,0.3,6.8,0.0
A231	Good	340	0.0,1.5,3.6,1.5,5.5,2.5

FIGURE 2.9
Data for roads on Figure 2.7.

NAME	QUALITY	TRAFF	X1	Y1	X2	Y2	X3	Y3	X4	Y4	X5	Y5
A23	Fair	1000	4.5	10.0	4.5	5.7	5.5	2.5	6.5	0.3	6.8	0.0
A231	Good	340	0.0	1.5	3.6	1.5	5.5	2.5				

FIGURE 2.10
Alternative method of storing locational information for lines.

ROADS

ID	NAME	QUALITY	TRAFF
1	A23	Fair	1000
2	A231	Good	340

ROADPTS

ID	SEQ	X	Y
1	1	4.5	10.0
1	2	4.5	5.7
1	3	5.5	2.5
1	4	6.5	0.3
1	5	6.8	0.0
2	1	0.0	1.5
2	2	3.6	1.5
2	3	5.5	2.5

FIGURE 2.11
Locational and attribute data for lines stored in two normalised tables.

defining that road. The two tables are linked by an ID value which is unique for each section of road.

So, it is possible to store the locational data for lines but only in a rather complex way. The situation would be similar for storing area boundaries. This is not our only problem with vector but before we come onto this, let us look at the storage of raster data.

Figure 2.12 shows part of the raster layer for buildings as it might look if it were placed in a relational table. On the face of it, a raster layer, which is made up of rows and columns of values, would seem to be easy to store in a relational table. However, the order of the rows and columns is absolutely crucial to the storage of raster data and this is not stored in the table in Figure 2.12.

The row which is labelled ROW0 is not seen by the relational system as the first row but as a row which happens to be called ROW0 – it could be called ERIC for all the difference it would make. The same is true for the column names. Again, the only way to store these data is to normalise it so that each pixel is stored in a single row as shown in Figure 2.13.

This will work, but we have now trebled the amount of information we are storing. We also have to store the geographical location of our pixels and

	COL0	COL1	COL2	COL3
ROW0	0	0	0	0
ROW1	0	0	1	0
ROW2	0	0	0	0
ROW3	0	0	0	0

FIGURE 2.12
A raster layer stored as a relational table.

PIXELS

ROW	COL	VALUE
0	0	0
0	1	0
0	2	0
0	3	0
1	0	0
1	1	0
1	2	1
1	3	0

FIGURE 2.13
Raster data stored in normalised form.

this would probably be done by having a second table called LAYERS with the minimum and maximum X and Y values for the overall layer, and an additional column in the PIXELS table indicating which layer these pixels belonged to.

The other difficulty with using a relational database to handle spatial data is that when GIS systems were first being developed, standard SQL had no way of performing quite basic GIS operations. There is no operation for finding the point which is nearest to somewhere or calculating the length of a line. If we know the length of a line, we can store it as an attribute and then use it in queries – but the relational system cannot calculate it for us.

2.4 Solutions to the Problems of Storing Spatial Data in RDBMS

Many of the early GIS systems solved these problems with storing and processing spatial data in an interesting way. The software writers realized that the attribute data for vector layers could be stored in a relational database so rather writing their own systems for handling these data, they simply used existing RDBMS systems. The locational data are stored and handled using specially written software which can handle its more complicated structure. Such systems are often referred to as georelational because the attributes are often held in a relational database, with the geographical or locational data being handled separately. The best known of these early systems is probably ARC/INFO, in which the attribute data were handled by a database package called INFO and the locational data were handled by a specially written software called ARC.

Having separated the location and the attribute data, such systems then have to make sure that they can linked back together again when necessary – for example, if a user selects a particular feature by querying the database, then to draw that feature on the screen, it will be necessary to retrieve its locational data. This is done by making sure that each feature has some sort of unique identifier which is stored with both the locational and attribute data.

There is a price to pay for this solution, however. It means that an organisation cannot hold its data in one system, but must use a separate system for its spatial data. It also sometimes meant that users had to learn two interfaces, one for doing spatial operations and a separate and often rather different one for working with the attribute data. Because of these issues, there have been constant developments in the link between GIS systems and database software. One approach has been to try and remove the limitations which make storing spatial data in a RDBMS difficult. This has involved relaxing some of the rules governing relational databases, to allow a table to contain one column which breaks the rule about atomicity and which can contain a list of *XY* pairs. In addition, the SQL language has been extended to allow spatial operations on these data. Initially, these developments were done by the vendors of specific database packages, but there has also been work to develop new versions of the international SQL standard which include these developments. This will be explored in a little more detail in the chapters on vector and raster data structures since the issues and the solutions are somewhat different for vector and raster data.

Another approach has been to use non-relational database systems which do not have the same rules and restrictions. One of the most interesting of these has arisen because of the growth of the Internet and the need to provide online access to large amounts of data. Large online retailers have found that relational systems are actually not well suited to their needs. The types of query which their customers make are simple and often consist of simply listing all items which match some basic criteria and listing them in price order. The key for these companies is not therefore to be able to construct complex databases with lots of joined tables but to operate reliably over the network 24 h a day. Among other things, this means having systems which can cope if parts of the network or individual computers fail. This has led to a move towards what are called NoSQL databases, which are much simpler than RDBMS, but which operate more reliably and efficiently over a network. As spatial data are being increasingly delivered across the Internet, there has also been a move to explore using NoSQL databases with spatial data. Finally, there are some users of GIS whose needs go beyond simple queries and who require complex spatial analysis and modelling tools which are simply not possible using a database system. For such users, a system based on some variant of the georelational model is still the best and will probably continue to be so for quite a long time.

Further Reading

The difficulties of storing spatial data in standard relational databases have been covered only briefly here. A more extensive discussion of the problems, and why georelational GIS systems were designed, is provided by Healey (1991) and Samet and Aref (1995) discuss some early attempts to integrate spatial data into relational databases. Shekhar and Chawla (2003) provide a good general introduction to databases in general, with a focus on their use for spatial data. DeCandia et al. (2007) provide a fascinating, and very readable account of why Amazon moved away from relational databases – their paper can be downloaded from http://www.read.seas.harvard.edu/~kohler/class/cs239-w08/decandia07dynamo.pdf. The homepage for the NoSQL 'movement' is at this web page: http://nosql-database.org/.

3

Vector Data Structures

In Chapter 2, we saw that the nature of spatial data means that it is difficult to store vector data in a database. In the days when GIS was first being developed, programmers were therefore forced to develop their own ways of storing and processing vector data. A common solution was to separate the attribute data as this could be handled using a standard database. This was not a universal solution by any means, but for the time being, we will concentrate on what was done to store the locational data for points, lines and areas.

3.1 Simple Storage of Vector Data

There is nothing intrinsically complicated about the locational element of vector data. If we return to our simple topographic map, we can see examples of the three types of vector feature: points, lines and areas.

As we saw in Chapter 2, the location of a point is simply a pair of coordinates, so to store the locational data for the three buildings in Figure 3.1, we need to be able to store three pairs of numbers and we need some way of telling which is which. This means the data we need to store are

```
1, 4.5, 2.5
2, 5.5, 2.5
3, 5.7, 2.4
```

The first number is an ID which identifies each point and allows it to be linked to its attributes. For the lines, each feature would now have multiple *XY* coordinate values:

```
1, 4.5,10.0,4.5,5.7,5.5,,2.5,6.5,0.3,6.8,0.0
2, 0.0,1.5,3.6,1.5,5.5,2.5
```

Again, the first number is an ID. For the polygons, the boundary of the polygon would simply be a line where the last point is the same as the first:

```
1, 4.5,10.0,4.5,5.7,5.5,,2.5,6.5,0.3,6.8,0.0,4.5,10.0
2, 0.0,1.5,3.6,1.5,5.5,2.5,0.0,1.5
```

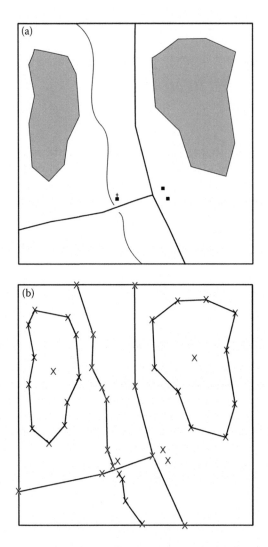

FIGURE 3.1
Topographic map (a) and features stored as vector points, lines and areas (b).

These three sets of data actually have the same basic structure. In each case, we have an ID for the feature and 1 or more points and so they can all be described by a single data structure of the form

```
ID, N, X1,Y1 . . . XN,YN
```

where

- ID is the unique identifier for the feature

```
<Placemark>
  <name>GeographyDepartment</name>
  <styleUrl>#msn_ylw-pushpin</styleUrl>
  <Polygon>
    <tessellate>1</tessellate>
    <outerBoundaryIs>
      <LinearRing>
        <coordinates>
          -1.489444250819044,53.38328878707264,0
          -1.489434522406109,53.38306879834025,0
          -1.488971012209974,53.38296744371188,0
          -1.488461329968353,53.38333587450911,0
          -1.48915992990775,53.38354883090607,0
          -1.489444250819044,53.38328878707264,0
        </coordinates>
      </LinearRing>
    </outerBoundaryIs>
  </Polygon>
</Placemark>
```

FIGURE 3.2
KML file for the storage of an area feature.

- X1, Y1 are the coordinates of the first point
- N is the number of points and XN, YN are the coordinates of the last point

In the case of points, N is always 1, that is there is only 1 coordinate pair. In the case of polygons XN = X1 and YN = Y1, that is the last point is the same as the first point. As an example of data stored in this way, Figure 3.2 shows how Google Earth stores the outline of a polygon in what is called Keyhole Markup Language (KML), which is a text format for storing geographical data.

The information in bold corresponds to the simple data structure described above. The ID is a name in this case, which is enclosed in the name tag

`<name>Geography Department</name>`

The number of coordinates is not given. Instead, the end of the list of the coordinates is indicated by the closing </coordinates> tag. As these data are going to be used in a global system, the X and Y coordinates are given in latitude and longitude as decimal degrees, for example

`-1.489444250819044, 53.38328878707264`

for the latitude of the first point. Decimal degrees are explained in a separate box if you are not familiar with them.

DECIMAL DEGREES

Since the earth is roughly spherical, global positions are recorded in terms of latitude and longitude, both of which are angular measurements. I am writing this book in my office at Sheffield University, which is at a latitude of 53° 23′ 00″ N and a longitude of 1° 29′ 21″ W. This way of expressing latitude and longitude is fine for anyone wishing to find this location on a map or a globe but it is not very convenient for use in GIS, which really needs a single number to express each coordinate. To see how these values can be converted to a single number let us start with latitude and ignore the fact that this is north of the Equator for the moment. The number of degrees is 53. Each degree is further subdivided into 60 minutes. This means that if the latitude were 53° 30′, then the position would be halfway between 53 and 54 and could be expressed as 53.5° because 30/60 is 0.5. So, to convert 53° 23′ to a decimal, we use the calculation 53 + 23/60, which gives 53.3833. The longitude is 1° 29′ 21″. We can convert 1° 29′ to 1.4833 but what about the 21 seconds? Each second is 1/60th of a minute so 21 seconds is 0.35 minutes. Each minute is 1/60th of a degree, so 0.35 minutes is 0.35/60 or 0.005833°. Dividing by 60 and by 60 again is the same as diving by 360 so 21 seconds is 21/360, which also gives 0.005833. This is rather a long explanation, but in fact the conversion to decimal degrees can be expressed as quite a simple formula:

$$DD = degrees + minutes/60 + seconds/360$$

Applying this to the original position gives

Latitude: 53° 23′ 00″ = 53.3833°
Longitude: 1° 29′ 21″ = 1.4891°

This just leaves the position relative to the Equator and the Greenwich Meridian. A convention is used in which position to the north and east of these two reference lines are positive, to the south and west negative. So the full conversion is

Latitude: 53 23′ 00″N = 53.3833 decimal degrees
Longitude: 1° 29′ 21″ W = –1.4891 decimal degrees

A very similar structure is used for lines and points in KML. The information provided by this data structure is sufficient to draw the features and so re-create the original map. If we have attribute information for each feature, we can use this to affect the way the feature is drawn. For example, if the roads are labelled to distinguish between major and minor roads, these can be drawn using different colours and line widths as is normal on topographic maps or motoring atlases. We can also do some simple calculations using the data in this format. To illustrate how this is done, we will use the pseudocode which was described in Chapter 1.

We saw in Chapter 1 that all programming languages have a means of storing lists of information in an array. In the sieve of Eratosthenes example, we just needed a list of whether each number from 1 to 100 was prime or not. To store the data for points, we need to store three pieces of information for each point – the ID, the X coordinate and the Y coordinate. This is not a problem, however – we simply add extra dimensions to the array:

```
POINTS[1..100,1..3]
```

This declares an array which has 100 rows each of which has three columns. To store the data for the first point, we could use

```
POINTS[1,1] = 1
POINTS[1,2] = 4.5
POINTS[1,3] = 2.5
```

It then becomes possible to use these coordinates to do calculations such as finding the distance between two of the points. To do this, we use one of the best-known expressions in mathematics, the Pythagoras' equation. This states that for any right-angled triangle, the square of the length of the hypotenuse is equal to the sum of the squares of the other two sides. With the sides labelled *a*, *b* and *c* as in Figure 3.3, this can be expressed as

$$a^2 + b^2 = c^2 \tag{3.1}$$

This equation can be rearranged to give *c* in terms of *a* and *b*

$$c = \sqrt{a^2 + b^2} \tag{3.2}$$

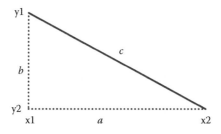

FIGURE 3.3
Illustration of how distance can be calculated using Pythagoras' theorem.

If you look at Figure 3.3, you can see that the length of the side labelled *a* can be calculated by taking *x*1 from *x*2. In the same way, *b* is the difference between *y*1 and *y*2. So, if we want to know the distance between these two points, the equation is

$$d = \sqrt{(x1 - x2)^2 + (y1 - y2)^2} \qquad (3.3)$$

If we wish to calculate the distance between point 1 and 25 then in pseudo-code this becomes

```
Distance = sqrt (power (POINTS[1,2]  -
POINTS[25,2],2) + power (POINTS[1,3]  - POINTS[25,3],2))
```

We can use the same basic equation to work out the length of a line by simply working out the distance between each pair of points and adding these up. In pseudocode

```
totaldistance = 0
For each point I from 1 to N-1
    distance = sqrt (power (Xi  - Xi + 1,2) + power (Yi  - Yi + 1,2))
    totaldistance = totaldistance + distance
```

In writing out this pseudocode, we have used a simple shorthand which uses the *X* and *Y* value of each point (*Xi*,*Yi*) and the next one along the line (*Xi* + 1, *Yi* + 1); but, when we come to implement this, we will have to decide how we are actually going to store all these *X* and *Y* values. With points, this is easy because each point has the same number of data values, so an array with three columns could be used. However, now each line will have a different number of data points – an ID and then a list of *XY* values. This was exactly what created the problem with storing these data in a relational database. In a programming language, however, there are several ways in which this could be handled. Let us start by assuming our language can only handle fixed size arrays of the kind we used for our points. We could store the data for the lines by using two arrays as shown in Figure 3.4.

The two arrays are as follows:

- An array called POINTS with a row for each point containing two values
  - The *X* coordinate for the point
  - The *Y* coordinate for the point
- An array called LINES with a row for each line containing three items of information
  - The ID of the line
  - The number of points in the line

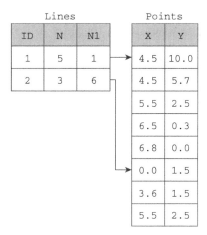

FIGURE 3.4
Storage of variable length lines in arrays of fixed dimensions.

- The number of the row in POINTS which contains the X and Y values for the first point in the line

Now, if we needed to work out the length of a line, we could use the code as shown above and we would find the appropriate X and Y values by finding where in the POINTS array the line begins and working down from there.

Many programming languages allow much more flexible arrays. In the Python language, for example we could combine all the information for each line and have a LINE array which we could declare as follows:

```
LINE[0] = [1,5,4.5,10.0,4.5,5.7,5.5,2.5,6.5,0.3,6.8,0.0]
LINE[1] = [2,3,0.0,1.5,3.6,1.5,5.5,2.5]
```

The first item in each element of LINE is the identifier, the second is the number of points and then the array is made up of the X and Y values. The first point always starts at the third position in LINE, so knowing the number of data points we can read the coordinates. Note that the rows of the LINE array are numbered starting at 0 rather than 1. This is also true for the elements in each row, so that the ID for the first line is expressed as LINE[0][0]. This may seem an odd way of doing things, but it makes some things much easier when writing programs and is quite common in computer programming and data structures.

We have seen that storing the X and Y coordinates for points and lines is relatively simple once the restrictions of the relational database are removed and that simply by storing the coordinates and the attributes we can draw the features and do some calculations. However, there are limits to what we can do when data are stored like this. At the moment, each line is stored as a completely separate feature with no connection with any other line.

Therefore, it is not possible to use the data in this format to plan routes along a network of lines. Even when using the lines to produce maps, it can be useful to know about the connections between lines because this means we can check whether the lines on the map join up properly. The simple data structure which we have considered so far is often referred to as 'spaghetti vector' because the lack of information about connections means the lines are like separate strands of spaghetti on a plate. Before going on to see how we can store the connections between lines, let us look at the simple storage of area features because this will introduce some other reasons why it might be useful to know about the connection between features.

We saw in Chapter 2 that areas are stored by using a line for the boundary. As with points and lines, we will probably wish to store attributes for our areas. With points and lines, we simply added a label to the point and line data stored in the database, but it makes less sense to add a label to the boundary of an area – we naturally think of attributes as being associated with the interior of an area rather than its boundary. Therefore, it is very common to store a centroid for each area, which is a point which is located inside the polygon as shown in Figure 3.1. The centroid can be defined by hand when the area is digitised, but many systems will automatically define one if this is not done. The centroid is commonly used to give a position for labels when drawing maps of area features and for this reason centroids are normally positioned near the centre of the area (as their name implies).

The use of centroids means that to store a single area in our GIS, we actually need to store two things – the line defining the boundary and the point defining the centroid. In fact, things become more complicated still because so far we have only dealt with the simplest type of area. Figure 3.5 shows the

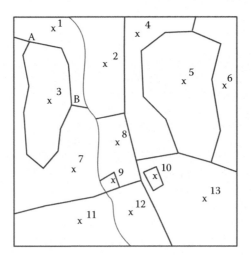

FIGURE 3.5
Landuse map – an example of multiple areas.

woodlands from Figure 3.1 as part of a landuse map rather than as separate areas. We now have a series of areas which neighbour each other, completely covering the area of the map. This type of map is very common – other examples are soil maps, geology maps and maps showing administrative areas such as countries, counties or districts.

Each area has a centroid, with an identifier associated with it, and this identifier is used as the link to a table containing the attributes for the areas (Figure 3.6).

We can still use the simple method of storing the areas but we will run into a number of problems. If we consider polygon 1 on Figure 3.5, we can see that this shares part of its boundary with one of the woodlands – between points A and B on the map. However, although we have already stored this part of the line in storing the woodland boundary, we have to store it again; otherwise there will be a gap in the boundary of the pasture. If we look at the whole map, we will see that the majority of the boundary lines lie between two areas, and will be stored twice in this way – the result is that we will store nearly twice as much data as necessary.

This is not the only problem. When we store the boundary, we choose a series of points along the line, and connect these by straight lines. When the same line is digitised a second time, slightly different points will be chosen, with the result shown in Figure 3.7.

Because the two lines do not coincide, there are small areas of overlap, and small gaps between the two areas. These mismatch areas are called sliver polygons because they are usually very small and thin.

There is a third problem with this method of storing area boundaries which arises if we wish to use our data for analysis rather than simply map drawing. We may wish to produce a new GIS layer which simply shows woodland

ID	Landuse
1	Pasture
2	Pasture
3	Woodland
4	Arable
5	Woodland
6	Pasture
7	Pasture
8	Pasture
9	Cemetary
10	Garden
11	Arable
12	Arable
13	Pasture

FIGURE 3.6
Attributes for landuse map.

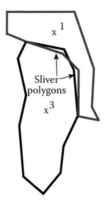

FIGURE 3.7
Sliver polygons as a result of digitising the same line twice.

and non-woodland areas. To do this, we will want to merge together all the polygons in which the landuse is not woodland – that is, to dissolve the boundaries between them resulting in a new layer looking like the one in Figure 3.8.

This operation is called a polygon dissolve, and is quite common in GIS analysis. However, it is difficult to do with the simple method of storing the area boundaries. If we consider polygon 1 in Figure 3.5, we do not know which part of the boundary has woodland on the other side, which means we would need to retain that part of the boundary, and which has some other landuse, meaning that we can drop that part of the line. In technical terms, we do not have any information about the contiguity of our polygons – which ones are neighbours to each other – and to store this information we need a different method of storing our area boundaries, which will be covered in the next section.

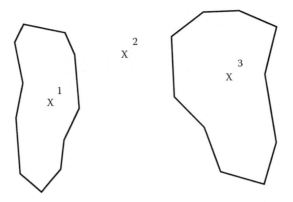

FIGURE 3.8
Map of woodland and non-woodland areas created using a polygon dissolve operation.

3.2 Topological Storage of Vector Data

To overcome the limitations of the simple method of storing polygons, GIS systems draw on ideas first developed in a branch of mathematics called topology. The link between topology and GIS will be explored in detail in the next section, but first let us look at the way in which vector data are stored in many GIS systems.

We saw in Figure 3.7 that when areas are stored using spaghetti vectors, we get problems caused by the fact that shared boundaries are stored twice. Figure 3.9 shows area 3 from Figure 3.6. It shares its boundary with two other areas – 1 and 7. Instead of storing the boundary of area 3 as a single line, an alternative is to split the boundary wherever it joins the boundaries of other areas as it does at the points marked A and B. This means that the boundary of area 3 is now stored as two separate lines as shown in Figure 3.10.

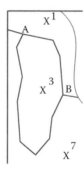

FIGURE 3.9
Part of the landuse map around area 3.

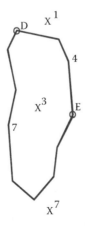

FIGURE 3.10
Boundary of area 3 stored as two connected lines.

Line	From	To	Area Left	Area Right	X1	Y1	X2	Y2	...
4	D	E	1	3					
7	E	D	7	3					

FIGURE 3.11
Storage of locational data for lines in Figure 3.9.

Each of the two new lines has been given its own identifier and the end of the lines has also been labelled because we will need to know how to join the lines together again to make up the boundary of area 3. The information which is stored for these two new lines in shown in Figure 3.11.

To draw the boundary of area 3, we simply take all the points from line 4 in sequence and then continue with line 7, which is effectively the same as when we had the boundary as a single line. This approach can be applied to the whole of the map, not just to individual areas. The full map is shown in Figure 3.12 and the locational data for areas 1 and 3 are shown in Figure 3.13.

One minor detail is what to do around the edges of the map. For instance, line 1 lies between polygon 1 and 'the outside world'. This is easily solved by assigning a special polygon number to the area outside the map – in this case it is 0.

If we need to draw the boundary of polygon 1, we would start by finding all the lines in the table in Figure 3.13 which have 1 as the AREA LEFT or AREA RIGHT as shown in Figure 3.14.

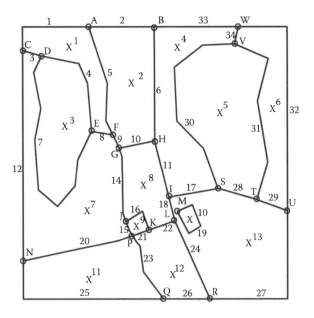

FIGURE 3.12
Landuse map as stored in a topological GIS.

Line	From	To	Area Left	Area Right	X1	Y1	X2	Y2	...
4	D	E	1	3					
7	E	D	7	3					
1	C	A	0	1					
3	D	C	7	1					
5	A	F	2	1					
8	F	E	7	1					

FIGURE 3.13
Locational data for areas 1 and 3 in Figure 3.11.

Line	From	To	Area Left	Area Right	X1	Y1	X2	Y2	...
4	D	E	1	3					
1	C	A	0	1					
3	D	C	7	1					
8	F	E	7	1					
5	A	F	2	1					

FIGURE 3.14
Data for area 1 from Figure 3.13.

The situation is now a little more complicated than before because we have five lines to join together. What is more, they do not all go in the same direction around polygon 1 as they did with polygon 3 as shown in Figure 3.15.

Three of the lines – 1, 3 and 5 – go around the polygon in a clockwise direction, but line 4 goes anticlockwise. We did not have this problem in the case of area 3, but as Figure 3.15 shows, any line which is shared between two

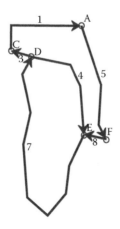

FIGURE 3.15
Lines around areas 1 and 3 showing the direction in which they are stored.

Line	From	To	Area_Left	Area_Right	Flipped	X1	Y1	X2	Y2	...
4	E	D	3	1	Y					
1	C	A	0	1	N					
3	D	C	7	1	N					
8	F	E	7	1	N					
5	A	F	2	1	N					

FIGURE 3.16
Locational data for area 1 adjusted to make all lines go in a clockwise direction.

areas will go in a clockwise direction around one area, but an anticlockwise direction around the other. This means that we need some way of dealing with this situation.

If all we wish to do is draw the boundary of polygon 1, we can do what we did before – draw each line in turn. Once all five are drawn, we will have our boundary. However, for some operations, we may need to be able to reconstruct the original polygon with the lines connecting in the correct order to form a single closed loop. All the information we need to do this is actually contained in the table in Figure 3.14. The first stage is to make all the lines go in the same direction. If a line goes in a clockwise direction around a polygon, then that polygon will be on the right-hand side. Line 4 currently goes from node D to node E and had polygon 1 on the left. However, if it went from E to D, then polygon 1 would be on the right and polygon 3 on the left. So, we can modify the table as shown in Figure 3.16.

Simply by swapping the FROM and TO and LEFT and RIGHT columns, we have changed the direction of the line. However, the *XY* values are still listed in the original sequence. We could swap them around too, but there is no need. Instead, we have added an extra column to the data called FLIPPED and we use this to record whether the direction of a line has been flipped. If it has, we know that when we read the *XY* values, we start at the end and work forwards. The next stage is to work out which order we need to draw the lines.

We start with the first line in the table. This is line 4 which starts at node E and ends at D. The next line in the sequence must therefore be the one that starts at node D. Inspection of the table shows that this is line 3. Now, it becomes clear why the nodes have been labelled because it allows us to easily find out which lines join each other.

So far, the description has been done in words, but it is also a simple matter to describe it as a series of procedures:

```
/* Check that 1 is on the right
/* Flip line if not
for each row
if Area_right != 1 then
```

```
    Temp = Area_left
    Area_left = Area_right
    Area_right = Temp
    Temp = From
    From = To
    To = Temp
    Flipped = Y
/* Draw lines in order. Note this code assumes that
/* each column in Figure 3.16 is stored in a separate array.
/* The notation FROM(1) therefore means the first cell in
/* the FROM column.
Current = 1
Start = FROM(Current)
ToNode = TO(Current)
do
  if Flipped (Current) == Y then
      Draw the current line from last point to first
  Else
      Draw the current line from first point to last
/* Find_next_line finds line that starts with current ToNode
/* and returns its LineNumber
  Current = find_next_line(ToNode)
  ToNode = TO(Current)
until ToNode == Start
```

The first half of this pseudocode checks whether the polygon we are interested in lies on the right-hand side of the segment. If it is not, the area and node IDs are swapped and the Flipped variable set to Y. The second half of the code starts with the first line in the list and makes a note of its start node in variable Start. On each loop, it then writes out the XY values or this line, and then calls procedure find_next_line. The details of this procedure are not shown, but we assume that it searches all the lines for one that begins with the 'To' node of the current line and returns the number of this line to the program. The program loops around until the next line is the one which we started with, at which point the program ends. The importance of being able to express this in pseudocode is that it shows that we have a data structure which can be processed by a computer.

This data structure solves the problems with the spaghetti vector structure outlined above. First, each part of the lines is only stored once, thus saving the problem of duplicating data and avoiding the problem of sliver polygons. Second, we now have information about the relationship between areas, which will allow us to do things like the polygon dissolves.

The same link and node structure can also be used for line data, where it is the connections between lines at the nodes, which is the important element – the left/right area identifiers are generally ignored. However, knowing about connections between lines means we can trace routes through networks, for example, a topic, which will be covered in Chapter 11.

This kind of data structure was taken up with great enthusiasm by GIS developers in the 1980s and 1990s. As often happens, when different companies develop an idea, different names emerged for the same thing. The points where the lines meet are almost always called nodes, but the sections of line in between are variously known as arcs, chains, segments, and links. However, the structure is often referred to as link-and-node, which is the terminology that will be used here. Such structures are often also referred to as topological because they have their origin in ideas developed in the mathematical subject of topology and this is what we will look at next.

3.3 So What Is Topology?

The study of relationships such as contiguity (whether objects are neighbours or not) is part of the mathematical subject of topology, which is concerned with those characteristics of objects which do not change when the object is deformed. For example, imagine the landuse map shown in Figure 3.5 printed on a sheet of thin rubber. If the rubber was stretched, then some properties of the areas would change, such as their size and shape. However, no amount of stretching could make polygon 3 border polygon 8 – this would involve cutting the sheet or folding it over on itself. Hence, the connection (or lack of it) between areas is a topological property of the map. Containment is another example of a topological property, since no amount of stretching will move centroid 3 outside its polygon.

One of the earliest people to study such properties was the Swiss mathematician Leonhard Euler (pronounced 'Oiler') and one of the classic problems he studied, the Konigsberg bridge problem, has a direct relevance to the use of topological ideas in GIS. In the town of Konigsberg, there were two islands in the Pregel river, which was connected to each other and to the banks of the river by seven bridges as shown in Figure 3.17.

The local people believed that it was impossible to plan a route which started and ended in the same place, but crossed every bridge only once. However, nobody could prove whether this was in fact correct. Euler realised that the problem had nothing to do with the distances or directions involved, but depended solely on the connections between places. He reformulated the problem by representing each of the land masses as points and the connections between them as lines as shown in Figure 3.18.

This representation is called a graph by mathematicians. The key to the problem is the number of lines which meet at any given vertex of the graph. If this is an even number, you can reach that vertex and leave by a different line. If it is an odd number, then eventually the pairs of entry/exit lines will be used up and there will be only one unused line joined to that vertex – that is, you can visit the vertex, but cannot leave again without using a line for

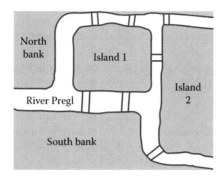

FIGURE 3.17
The Konigsberg bridge problem.

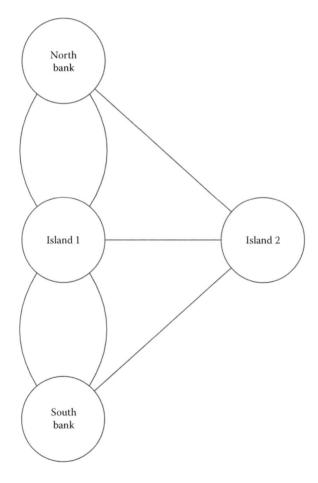

FIGURE 3.18
Graph of the Konigsberg bridge problem.

the second time. Therefore, it is only possible to make a round trip walk if all vertices have an even numbers of lines, or if there are just two vertices at which an odd number of lines meet (which will have to be the start and end points of the route). In the case of the Konigsberg bridges, neither condition is satisfied, proving that the round trip cannot be made and that the locals were right.

Another mathematician, Henri Poincaré, realised that graph theory could be applied to maps in general, and his ideas were used by the staff at the U.S. Bureau of the Census to help in processing the data for the 1980 census. A key part of processing the census data was in handling the map of the street network of the United States. This indicated where each address was located and in which block it fell. To automate the processing of census data, it was necessary to have an accurate database which indicated which block each address was in. Compiling such a database was an enormous task in which errors were bound to be made, and so some way of checking the accuracy of the results was needed.

The map in Figure 3.19 is a fictitious example of part of the street network in an American city. It can be seen that each block is surrounded by sections of street which meet at junctions. If we treat each street intersection as a vertex, we can regard the street network as a mathematical graph. What is more, if we consider the part of the graph which surrounds an individual block (e.g., Block 5 in the above map), it is clear that this will form a single connected circuit.

However, we can also use a graph to represent the relationship between the blocks. First, we represent each block by a single point as shown with blocks 5, 6, 8 and 9 in Figure 3.19. The points are then joined by lines if

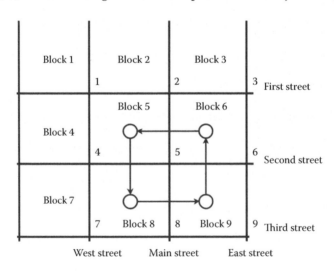

FIGURE 3.19
Fictitious city blocks illustrating Poincaré's dual graph model of maps.

they lie on the either side of a street section and we have a graph similar to the one constructed by Euler for the Konigsberg bridge problem. If we take the portion of this graph which surrounds a single street intersection (like the example shown in Figure 3.19, which surrounds node 5), then this should form a single connected circuit as with the first graph.

We can therefore construct two graphs – one based on the streets surrounding a block and one on the blocks surrounding a street intersection – and it was this model of maps as paired graphs which came from the work of Poincaré. Mathematically, the two graphs are exactly the same, since both will consist of a single closed loop. If we can automatically create these graphs from our street map, and check that they do form closed loops, we will have a way of checking the data. This is exactly what the staff at the Bureau of the Census managed to do when they developed a data structure called DIME, which will be explained in the next section.

3.4 And How Does It Help? The Example of DIME

So far, we have considered these graphs by drawing them on the original map. However, computers cannot 'see' maps, so we must devise a method of storing the data in a format which allows the computer to construct and test the graph around any block or junction. The system which was devised was called DIME – dual independent map encoding – and was based on a data structure in which one record was created for each street segment – the part of a street between two junctions. For the map in Figure 3.19, a portion of the DIME file might be as shown in Figure 3.20.

The alert reader will notice that this is very similar to the topological data structure described earlier. In Section 3.3, I described how this information could be used to reconstruct the boundary around a polygon. However, this

Seg#	From	To	Block Left	Block Right	Street Name
1	1	2	2	5	First
2	2	3	3	6	First
3	4	5	5	8	Second
4	5	6	6	9	Second
5	4	1	4	5	West
6	7	4	7	8	West
7	5	2	5	6	Main
8	8	5	8	9	Main

FIGURE 3.20
DIME data structure for fictitious city map.

assumes that the information provided is correct and that it does describe a closed boundary around the area. The interesting thing is that the same information can be used to check whether this is in fact the case, and this is how the DIME system was used.

If we look at block 5 in Figure 3.19, it is very simple for us to see that it is surrounded by four connected street segments. In order to check this on the computer using the data structure, we first find those records in the DIME file which have 5 on either their left or right. Since the DIME file is simply a table, this could be done using a standard database query, or by writing a program to read each record and select those in which the block left or block right value was 5. In this case, we will find segments 1, 3, 7 and 9. We then need to check that these segments form a closed loop, and this is most easily done if they all go in the same direction around the block. If they all have block 5 on their right-hand side, this means they will form a clockwise loop. At the moment, only segments 1 and 7 have 5 as their right-hand block – however, to change the direction of the other two, all we need to do is switch the left and right blocks and the to and from nodes to produce the records shown in Figure 3.21.

We can now start at any segment and try and trace round the loop. We start at segment 1, noting that our starting point is node 1. The end of this segment is node 2, so we look for a segment which starts at node 2. We find segment 9, which ends in node 5. This leads us to segment 3 which ends in node 4, and this leads us to segment 7, which ends in node 1, our starting point. If for any reason we cannot complete this loop, we know there is an error in the data, such as a segment completely missed out, or one in which the block numbers or node numbers were wrong. For example, if segment 2 had 5 as its right block instead of 6, this would create three errors:

- Block 3 would not be correct because of the 'missing' segment.
- Block 6 would not be correct because of the 'missing' segment.
- Block 5 would close but would have a segment unused.

Seg#	From	To	Block Left	Block Right	Street Name
1	1	2	2	5	First
3	5	4	8	5	Second
5	4	1	4	5	West
7	2	5	6	5	Main

FIGURE 3.21
Records from DIME file relating to block 5. All records have been modified so that block 5 is on the right-hand side of the street.

Seg#	From	To	Block Left	Block Right	Street Name
3	4	5	5	8	Second
4	5	6	6	9	Second
7	5	2	5	6	Main
8	8	5	8	9	Main

FIGURE 3.22
Records from DIME file relating to junction 5.

This checking process can also be carried out using the part of the graph which surrounds each street junction. If we consider junction 5, then we can identify the segments which meet at this point because they will have 5 as either their start or end node (Figure 3.22).

We then adjust these so that they all end at node 5 as shown in Figure 3.23.

Now, if we start at a segment, then the block on the right of that segment must be the left-hand block of one other segment, which in turn will share its right-hand block with one other segment, until we work our way around to the starting point.

In both block and junction checking, we are using the left/right and from/to information to trace around the topological graph. Since we know from mathematical theory that there must be one closed graph in each case, if we fail to find this, we know we have an error in the data.

Notice that the original DIME file did not contain any geographical coordinates. The geographical referencing was done via addresses (Figure 3.24) since the original file had extra fields which have not been shown so far, which indicated the range of addresses on either side of the segment. This allowed any address to be matched with its block number (and census tract number), and also allowed summary statistics to be produced for any block by aggregating data for all the addresses.

We have now seen how ideas from topology lead to the DIME data structure. Since this was intended simply for handling data for streets, the segments in the DIME file were all straight – if a street curved, it was simply broken up into segments. To develop this into a more general-purpose data

Seg#	From	To	Block Left	Block Right	Street Name
3	4	5	5	8	Second
4	6	5	9	6	Second
7	2	5	6	5	Main
8	8	5	8	9	Main

FIGURE 3.23
Records from DIME file relating to junction 5 modified so that the street ends at junction 5.

Seg#	From	To	Block Left	Block Right	Street Name	Left Address Low	Left Address High	Right Address Low	Right Address High
1	1	2	2	5	First	12	24	13	25
2	2	3	3	6	First	26	36	27	37

FIGURE 3.24
Storage of address information in the DIME data structure.

structure, it was necessary to allow the lines between the nodes to take on any shape, as described by a set of *XY* coordinates – in this way, we reach the link and node data structure which was described earlier.

3.5 More on Topological Data Structures

So far, we have concentrated on what information is essential to store in order to be able to process vector data. However, in the design of data structures, it is sometimes the case that additional information is stored because this will make it easier or faster to perform certain operations. To illustrate the point, we will look at the use of the link and node data structure for storing line data.

Figure 3.25 shows part of a road network consisting of five roads connecting four nodes. The topological data for this network are shown in Figure 3.26.

Note that in this case, since we are dealing with lines, there is no information relating to the areas on the either side of the line. The query we are going to consider is

```
Is there a direct route from A to D?
```

This is an extremely simple query, but this will allow us to focus on the data structure. Later in the book, we will consider more realistic operations on networks.

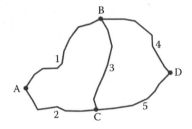

FIGURE 3.25
Road network stored in link and node format.

Line	From	To
1	A	B
2	A	C
3	C	B
4	B	D
5	C	D

FIGURE 3.26
Topological data for road network.

To answer this query, all we need to do is see whether a link exists which has A at one end and D at the other. Note again that although we can answer this query in an instant by looking at Figure 3.25, the computer can only 'see' the table in Figure 3.26, so we need a method which can work with these data and which does not require knowledge or intuition. In this case, the procedure is simple – we go through the links looking for one with A at one end and D at the other. In pseudocode

```
Direct = FALSE
For each link
    IF From == A and To == D OR From == D and To == A
    Direct = TRUE
```

We set a variable to false and then look at each link in turn. If none satisfies, the condition direct remains as false; otherwise, it is set to true.

Now, imagine the situation if we are dealing with a real network which will have considerably more links than in this simple example. For example, the OpenStreetMap project (www.openstreetmap.org) has roughly 250,000 road links covering the United Kingdom. Areas such as Europe or the United States will contain considerably more. To answer our query now requires us to look at every single link to find just one. This would be considerably speeded up if instead we could look at the two nodes in question, A and D, and see whether a link connects them. This would be possible if we construct a node table as shown in Figure 3.27.

Node	Links
A	1,2
B	1,3,4
C	2,3,5
D	4,5

FIGURE 3.27
Node table for network in Figure 3.24.

Now, our pseudocode becomes

```
Direct = FALSE
Find record for node A in nodes table
For each link
    IF To == D OR From == D
    Direct = TRUE
```

Now, instead of searching all the links, we only need to search those connected to node A, which in most roads networks will typically be 3 or 4. Of course, there is a cost associated with constructing the node table. It is not a difficult thing to do. You would simply set up an array with one row for each node and then run through the link table adding the link identifiers into the node table in the rows corresponding to the start and end nodes. The important point is that this takes about as long to do as answering the actual query, so if it is only going to be done once it is not worth it. However, if the query is going to be run many times, then the time taken creating the node table will soon be saved.

This illustrates the fact that data structures tend to be designed according to two principles. First, they need to store the information which makes certain operations possible. For instance, the spaghetti vector structure makes it possible to draw a map of the roads but not possible to plan routes which requires a topological structure. Second, they can be designed to store additional information to make certain operations quicker as with the additional node table.

The same principle can be applied to area data. We saw earlier that in reconstructing the boundary of a polygon from a set of links stored in a link and node data structure, we always have to change the direction of some of the links so that they all run in the same direction around the area. In Figure 3.16, this fact was recorded by adding an extra column to the data structure. However, this means we would have to reset the information in this column every time we wished to consider a different polygon. An alternative approach is to work out which links form the boundary of each polygon and whether they need to be flipped or not and store this information. This is the method which was used in the storage of what were called coverages in the ARC/INFO GIS.

Figure 3.28 shows how the information for polygons 1 and 3 in Figure 3.15 might be stored in a vector topological system. In an ARC/INFO coverage, this spatial information would be stored as three separate files as shown in Figure 3.29. The topological data are split into two separate files, called the arc attribute table and the arc definition file. This is because the arc attribute table is in normal form and can be stored in the relational database. Each row has five columns – the number of the arc, the from and to node and the left and right areas. All the cells are atomic and there is no sequence to either the rows or columns. The ARC file contains the *XY* coordinates and is essentially the same as the spaghetti data structure described in Section 3.2. Finally, there was an additional file called the polygon definition file which recorded

Line	From	To	Area Left	Area Right	X1	Y1	X2	Y2	
4	D	E	1	3					
7	E	D	7	3					
1	C	A	0	1					
3	D	C	7	1					
5	A	F	2	1					
8	F	E	7	1					

FIGURE 3.28
Storage of boundaries of polygons 1 and 3 in a topological system.

which arcs formed the boundary of each polygon. The arcs are listed in order around the polygon, and when an arc needs to be flipped, its ID is given as a negative number. Finally, there was a fourth file which is not shown in Figure 3.29 which contained the attributes for each polygon, also stored in the database.

Line	From	To	Area Left	Area Right
4	D	E	1	3
7	E	D	7	3
1	C	A	0	1
3	D	C	7	1
5	A	F	2	1
8	F	E	7	1

Arc Attribute Table

Line	Coordinates
4	X1, Y1, X2, Y2, X3, Y3, X4, Y4
7	X1, Y1, X2, Y2, X3, Y3, X4, Y4, X5, Y5, X6, Y6, X7, Y7, X8, Y8
1	X1, Y1, X2, Y2, X3, Y3
3	X1, Y1, X2, Y2
5	X1, Y1, X2, Y2, X3, Y3, X4, Y4
8	X1, Y1, X2, Y2

Arc Definition File

Area	Arcs
1	1, 5, 8, −4, 3
3	4, 7

Polygon Definition File

FIGURE 3.29
Storage of area data in an ARC/INFO coverage.

3.6 And a Return to Simple Data Structures

We have seen that topological data structures store much more information about vector data and that this makes some types of analysis possible.
However, the cost of this is that the data structures become very complex as
shown in Figure 3.29 – this is a long way from the simplicity of a spaghetti
vector file such as the KML data shown in Figure 3.2. In addition, the operations for which the topological information is really vital, such as polygon
overlay and polygon dissolve, are relatively rare. Most GIS usage is far more
simple than this and involves map drawing and relatively simple queries.
This is even more the case with the use of spatial data on the Internet in
which map drawing and spatial searches (e.g., 'which restaurants are near
this point') predominate. In these situations, the extra data and complexity
involved in topological data structures are simply not warranted. This has
led some people to question whether topological data structures are necessarily always the best. Topological data structures are normally considered to
have three advantages compared with non-topological structures (Theobald
2001, Louwsma et al. 2003):

1. Storing topological information explicitly is necessary for some
   operations.
2. Storing topology makes it possible to check for errors in the data
   such as lines which do not meet.
3. By reducing the storage of duplicate boundaries, file sizes are smaller.

Theobald (2001) shows that GIS developers have increasingly questioned
these assumed advantages. It has been found that the operations which
require topology to be stored are performed so infrequently that it is often
better to calculate the necessary topological information when it is needed
and throw it away afterwards – calculating 'on-the-fly' as it is called. The
point about error checking dates from a time when data capture was done
by digitizing all the points and lines and then sorting them out at a later
stage. With modern computers and graphical interfaces, data are corrected
as they are created, with new features being automatically matched and
clipped to existing ones. Finally, the overhead associated with storing the
additional topological information means that the saving in file size may
be small.

Set against the proposed advantages, there is one large disadvantage of
topological data structures, which is that the information relating to a single
feature is not held in one place, but is scattered between numerous files and
tables as shown in Figure 3.29. This has numerous consequences. It means
that relatively simple operations, such as drawing a polygon, are slowed
down by having to reassemble the polygon from its constituent parts. This

has become an increasing issue as the size of spatial datasets has increased. In the early days of development, datasets would generally relate to small geographical areas and the system was operating on a single computer. Now, datasets can be global in coverage and be accessed by numerous computers across a network. As was the case with databases, in this situation, a simple data structure can produce much faster results. Second, it is difficult to transfer a complex data structure between users or systems, whereas the simpler spaghetti structure can easily be saved in a text format as shown by the example of KML (Figure 3.2). Finally, we saw in Chapter 2 that there are good reasons for wishing to be able to handle spatial data in standard database systems and this is greatly simplified if each geographical feature is represented as a single entity in a relational table instead of as a complex set of relationships between tables.

All of these issues have led to a move back towards the use of non-topological data structures for storing vector data. The desire for a simple data structure to make simple applications faster is what led ESRI (1998) to develop the shape file and led to the development of KML, which is used by Google Earth and Bing maps (Figure 3.2). KML allows users to produce placemarkers, indicating features of interest, and to link these to other information such as photographs or web addresses. This requires a vector data structure in that it is necessary to store the coordinates of the feature and the attributes, but it does not require any information about the relationship of a feature with any other. This means that a basic spaghetti vector structure is sufficient. Storing the boundary of a polygon in a link and node structure would simply complicate things.

The other motivation for looking at simple data structures has been a continued desire to be able to handle spatial data alongside other information inside standard database systems. This can be seen as part of a trend over recent years for spatial data handling to move away from being a specialist activity and becoming regarded as part of mainstream computing. The initial work to adapt RDBMS systems to handle spatial data often made use of what is known as a BLOB, which is variously translated as binary large object or basic large object. Relational databases are excellent at retrieving information based on complex criteria; but, as we saw in Chapter 2, the rules which in part make this possible restrict the type of data which they can handle. The idea of BLOB is to take all the messy information, which the database cannot process, and pack it away in a single additional column in each row. The database can retrieve this along with the other information and can pass it on to another program but it cannot necessarily do anything with it. If we return to the theatre database from Chapter 2, then our table of plays (Figure 3.30) could contain the poster for each play as a digitised image.

This could be displayed on screen when a customer requested information about plays being performed. However, the customer would not be able to select plays based on the images. For example, if you knew that a particular

PLAY	AUTHOR	DIRECTOR	POSTER
Twelfth Night	William Shakespeare	Trevor Nunn	Twelfth_night.jpg
Hedda Gabler	Henrik Ibsen	Jonathan Miller	Hedda_gabler.jpg

FIGURE 3.30
The PLAYS table from Chapter 2 with the publicity poster stored as a jpg in a BLOB column.

actor was in a production, you could not find out which one by asking the database to select the poster with their face on it.

This is not an entirely satisfactory solution since you ideally want the database to at least be able to return records based on spatial criteria (e.g., all theatres within London) even if it cannot do everything that a specialist GIS package can. Work has therefore continued to explore ways to store vector data inside a relational database and extend SQL to be able to perform some spatial operations on the data. This culminated in the production of a new version of the SQL standard which allowed SQL to process a range of data types, including spatial data. This standard makes use of a model of vector data which is somewhere between a spaghetti model and a full topological model. The main spatial objects are shown in Figure 3.31.

In this model, polygons are stored with their boundaries complete, not as sets of linked lines. There is a multipolygon type (which is not illustrated), but this is simply a collection of separate objects with no explicit connection between them. However, polygons can contain other polygons within them and both lines and polygon boundaries have a direction and a start and end node. Even though a full topological structure is not used, the extensions to SQL do allow for topological queries such as overlap and containment.

The thing to take away from this chapter is that there is no single way of storing vector data in GIS, but a variety of ways. This is partly a matter of change over time. The topological structures were developed to try and address the shortcomings of the early spaghetti data structures but then found to introduce complexities which were sometimes unhelpful and so

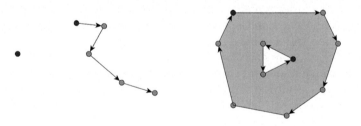

FIGURE 3.31
ISO and OGC simple feature model – point, linestring and polygon. Dark grey circles indicate nodes, light grey circle vertices.

different solutions were sought. It is also partly a reflection of the fact that the range of uses for spatial data is very large. The person who simply needs to produce maps will find a simple spaghetti structure will suffice but if you need to plan a route along a road network you are going to need some information about topology.

Further Reading

A number of authors have discussed the data structures used to handle vector data. Peucker and Chrisman (1975) is a standard reference here, as are the papers by Peuquet on both vector and raster (1984, 1981a, 1981b). More recent work includes a discussion by Keating et al. (1987).

The background to the DIME data structure is given in a paper by the US Bureau of the Census (1990) which is in the collection of papers by Peuquet and Marble (1990). The paper by White (1984) is a very readable summary of the importance of topological structuring in handling vector data. Worboys (1995) and Jones (1997) both have good sections on vector data structures, giving detailed examples which go beyond the material covered in this book. The NCGIA core curriculum (http://www.geog.ubc.ca/courses/klink/gis.notes/ncgia/toc.html) Unit 30 covers the structures described here, while Unit 31 describes chain codes, which are an alternative data which are a very compact way of representing vector lines. Nievergelt and Widmayer (1997) provide an excellent review of why data structures are needed and why specialised ones have had to be developed for spatial data. Theobald (2001) has a very good discussion of the reason why GIS writers have moved back to using non-topological data structures. The data structures used by the ESRI GIS products are often used as examples, partly because the software is so widely used and partly because the structures are well documented. For instance, ESRI (2008) is a white paper describing one of their non-topological structures, the shape file. Louwsma et al. (2003) and Baars (2003) discuss the relative merits of storing topology or calculating it as needed by looking at a range of commercial systems. The OGC standards for spatial data (OGC 2006) are fully documented on the OGC website (http://www.opengeospatial.org/standards/sfa), but in detailed documents which are a bit overwhelming for the non-specialist. Fortunately, Stolze (2003) provides a readable summary and the Microsoft Developer Network have a page which explains how the standards are implemented in SQL/Server (MSDN 2012). The algorithms and data structures used for vector spatial data have much in common with those used in vector computer graphics. For example, to window in on a detailed line drawing, such as an architect's plan, it is necessary to decide which lines lie within the window, which outside, and which cross the window boundary – in other words, a classic line intersection problem. The standard reference work on computer graphics is Foley et al. (1990), but Bowyer and Woodwark (1983) is an excellent, simple introduction to some of the fundamental algorithms. The frequently asked questions (FAQ) file from the comp.graphics.algorithms newsgroup also has some useful information and is archived at http://www.faqs.org/faqs/graphics/algorithms-faq/.

4

Vector Algorithms for Lines

Because points, lines and areas are so different from one another, the type of queries they are used for and hence the algorithms developed for them are sometimes different. Some queries are just not meaningful for all vector types. For example, polygons can contain points, but points cannot contain polygons, hence a point in a polygon test is useful, but a polygon in point test is not. Even when queries are meaningful for all types, they may need to work rather differently. For instance, it is possible to measure the distance between any two features. If both are points, this is simple, but if they are areas, then there are several possible distances: the distance between the nearest points on the boundaries, and the distance between the centroids.

The algorithms which are described in this chapter and the next one have been chosen because they are among the most fundamental vector algorithms. They are not only useful in their own right but also often underpin other more complicated algorithms. They also illustrate some important characteristics of good algorithm design, and the importance of the topological data structure introduced in Chapter 3.

4.1 Simple Line Intersection Algorithm

We have already seen (in Chapter 3) how a vector GIS can store a line feature by storing the coordinates of a series of points along the line. This section and the next two will deal with how we can use this representation to decide whether two lines intersect or not. This sort of question may be useful in its own right – for example, to see whether a proposed new route crosses any existing routes or rivers, which would necessitate building bridges. Working out whether lines intersect is also a key part of many other GIS operations, such as polygon overlay and buffering.

The problem then is to decide whether two lines, such as those shown in Figure 4.1, cross each other. This might seem a trivial question to answer – most people reading this could give the correct answer instantaneously and with no conscious thought. However, as we shall see, to get a computer to give the same answer is not at all an easy matter.

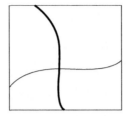

FIGURE 4.1
Line intersection problem.

Let us start by simplifying the problem. We know that each of the lines in Figure 4.1 is represented by a series of short straight segments between pairs of points. Let us first see how the computer decides whether two of these straight segments intersect or not, by considering the example of the two segments in Figure 4.2.

Again, this is a trivial question for a human being to answer but far more difficult for a computer. The first reason for this is that we can see both lines whereas all the computer can 'see' is the eight numbers representing the start and end points of each line segment as shown in Figure 4.3.

The question that the computer has to answer therefore is 'Do the two line segments with these start and end coordinates intersect?' The answer to this question is much less obvious even to a human observer, and clearly we need a method to work it out. What we need is a set of steps we can take which will allow us to take these eight coordinate values and work out whether the line segments they represent intersect – in other words, an algorithm. For our first attempt at an algorithm, we will need to invoke the aid of some elementary mathematics.

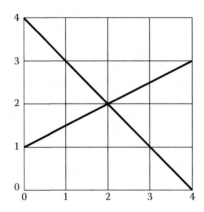

FIGURE 4.2
Testing for the intersection of two line segment intersections.

		X	Y
Line 1:	Start	0	1
	End	4	3
Line 2:	Start	0	4
	End	4	0

FIGURE 4.3
Coordinates of line segments.

For simplicity, the term 'line' will be used to refer to the line segments. We will return to the issue of lines made up of a series of segments in due course. One equation commonly used to represent straight lines is

$$y = a + b \cdot x$$

(the dot between b and x means 'multiply'). For the two line segments in Figure 4.2, the equations are

$$\text{Line 1: } Y1 = 1 + 0.5 \cdot X1$$

$$\text{Line 2: } Y2 = 4 - X2$$

This means that for any value of X, we can calculate Y – that is, if we know how far along the X axis we are, we can also work out how high up the Y axis the line is at that point. For line 1, therefore, when X is 3, Y will be $1 + 3 \times 0.5$, which is 2.5.

The important point about this is that if the two lines cross, there will be a point where the X and Y values for the lines are the same, that is the value of Y on line 1 will be equal to Y on line 2. In mathematical notation

$$Y1 = Y2 \quad \text{and} \quad X1 = X2$$

If we call these common values X and Y, then our two equations for the lines become

$$Y = 1 + 0.5 \cdot X$$

$$Y = 4 - X$$

which means that

$$1 + 0.5 \cdot X = 4 - X$$

There is now only one quantity in this equation which we do not know, that is X, and we can rearrange to obtain the following:

$$0.5 \cdot X + X = 4 - 1$$
$$1.5 \cdot X = 3$$
$$X = 3/1.5$$
$$X = 2$$

This means the two lines cross at the point where $X = 2$ (which agrees with the diagram, which is reassuring!). To find the Y coordinate for this point, we can use either of our equations for our lines:

$$Y1 = 1 + 0.5\ X$$
$$Y1 = 1 + 0.5*2$$
$$Y1 = 2$$

Therefore, the two lines cross at the point 2,2.

However, this has only showed the solution for two particular lines – to be an algorithm, our method must be able to cope with any pair of lines. To do this, we can generalise as follows:

Given two lines with end points (XS1,YS1), (XE1,YE1) and (XS2,YS2), (XE2,YE2), find the intersection point.

1. First, we must work out the equation of the straight lines connecting each pair of coordinates. The details of how this is done are given in Box 4.1 and this gives us values of A and B for line 1 and line 2.

2. Given the values for $A1$ and $B1$ and $A2$ and $B2$, we know that at the point of intersection

$$Y1 = Y2$$

Hence, $A1 + B1 \cdot X1 = A2 + B2 \cdot X2$.

However, we also know that where the lines meet the two X values will also be the same. Denoting this value by XP

$$A1 + B1 \cdot XP = A2 + B2 \cdot XP$$

We can now rearrange this equation to give

$$B1 \cdot XP - B2 \cdot XP = A2 - A1$$
$$XP(B1 - B2) = A2 - A1$$
$$XP = (A2 - A1)/(B1 - B2)$$

This gives us our value for X at the point where the lines meet. The value for Y comes from either of the equations of our lines:

$$YP = A1 + B1*XP$$

We now have an algorithm which will work out where the two lines intersect. However, this does not solve our problem as we shall see.

CALCULATING THE EQUATION OF A LINE

Given the coordinates of any two points on a straight line, it is possible to work out the equation of the line. To understand how to do this, let us first look at what the equation of the line means. The general equation for a line is

$$Y = A + B \cdot X$$

and the equation of the line in the diagram is

$$Y = 1 + 0.5X$$

that is $A = 1$, and $B = 0.5$.

B is the slope of the line – for each unit in X, it tells us how much the line rises or falls. In this example, Y rises by 2 units over an X distance of 4 units – hence, the slope is 2/4 or 0.5. Therefore, to calculate B, all we need is to look at the change in Y value between any two points and divide by the number of X units between the points.

If we have the X and Y coordinates of two points on the line, this is simply the difference in Y values divided by the difference in X or

$$B = \frac{Y2 - Y1}{X2 - X1}$$

In the case of this line

$$B = \frac{3 - 1}{4 - 0} = \frac{2}{4} = 0.5$$

To see what A means, consider what happens to our equation when $X = 0$. In this case

$$Y = A + B \cdot 0$$

Since, anything multiplied by 0 equals 0; this means

$$Y = A$$

A is the value of Y when X is 0. Put in the other way, it tells us how far up (or down) the Y axis the line is when X is 0.

In the present case, we already know the answer because one of our coordinates has a zero X value. But, we can work A out from the other point. We know the Y value (3) when X is 4. We also know how Y will change by as we change X – if we reduce X by 1 unit, Y will fall by 0.5. If we reduce X by 4 units (i.e., so that it is 0), Y will fall by 4×0.5 or 2, so giving a value of 1.

In mathematical terms, this means that

$$A = Y - X \cdot B$$

where X and Y can be taken from either of our points.

4.2 Why the Simple Line Intersection Algorithm Would Not Work: A Better Algorithm

To see what is missing from our algorithm, consider the two sets of lines in Figure 4.4. Figure 4.4a shows the lines used in the previous section while in Figure 4.4b one of the lines is somewhat shorter so that the two lines do not intersect. However, if you take the coordinates for this shorter line and use them to work out the equation for the line, they will give exactly the same result as for the case on the left: $Y = 4 - X$. The reason is that the equation describes the infinite line which passes through the two points – it says nothing about how long the line is. Since two infinite lines will always intersect somewhere, unless they are parallel, testing to see whether they intersect or not will not tell us whether the parts of the lines we are dealing with intersect or not.

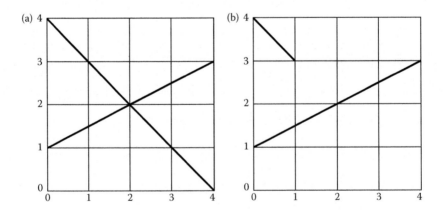

FIGURE 4.4
Line intersection problem – each pair of lines is defined by the same mathematical equation.

To finish the algorithm, we need to test to see whether the point at which the infinite lines intersect actually lies on the lines (Figure 4.4a) or lies beyond the end of the lines (Figure 4.4b).

To do this, we can look at the X value of the intersection point, and see whether it lies between the X values for the end of the line. From the previous calculations, we know that these two lines intersect at (2,2). Figure 4.4a, line 2 runs from $X = 0$ to $X = 4$; thus, since 2 lies between these values, the intersection point falls on line 1. In Figure 4.4b, line 2 runs from $X = 0$ to $X = 1$, the intersection point lies beyond the line and so the lines do not intersect.

Note that it is not enough to test one line. If we repeat the above test for the case of line 1, this will give the same answer in both cases – the intersection point does indeed fall on line 1 in both cases. However, as we can see, in the second case, the lines do not intersect – but, we can only be sure that this is the case by testing both lines.

The test which is used is to calculate the following:

$$(x1 - xp){\cdot}(xp - x2)$$

If this expression gives a result which is greater than or equal to 0, xp lies in between $x1$ and $x2$. If we do this test for the X and Y values of both lines, and all four produce values which are positive or zero, then the two lines intersect.

So, now do we have our algorithm? Well, yes and no! A proper algorithm must be able to deal with all situations without giving the wrong answer or causing problems, but we have so far overlooked two special cases which will trip our current algorithm up – the situation when our lines are parallel, or when either of them is vertical.

If the lines are parallel, then we know that they cannot intersect – so, surely our algorithm will simply report no intersection? To see why it will not, it is useful to look at the algorithm in pseudocode up to the point where we test for intersection:

```
/* Given end points of lines:
/*
/* Line 1: xs1,ys1
/*         xe1,ye1
/* Line 2: xs2,ys2
/*         xe2,ye2
/* Calculate A and B for equation of each line
/* Y = A + B*X
/* Note that the sign / is used for division
b1 = (ye1-ys1)/(xe1-xs1)
a1 = ys1-b1*xs1
b2 = (ye2-ys2)/(xe2-xs2)
a2 = ys2-b2*xs2
/* Calculate the X value of the intersection point xp
```

```
/* and the Y value yp
xp = (a2 - a1) / (b1 - b2)
yp = a1 + b1*xp
```

To understand why parallel lines give a problem, we need to remember that in the equation

$$y = a + b \cdot x$$

the value b represents the slope of the line. So, in Figure 4.4a, the line $Y = 1 + 0.5$ X has a slope of 0.5 – each step of 1 along the X axis results in a rise of the line by 0.5 units. The problem comes when we try and calculate the value of xp using the equation:

$$xp = (a2 - a1)/(b1 - b2)$$

If the two lines are parallel, they will have the same slope, so that $b1 - b2$ will equal zero. In mathematics, dividing by zero conventionally gives the answer of infinity. On the computer, it causes a program to crash, usually with a message referring to 'attempt to divide by zero'. The only way to prevent this is by checking the values of $b1$ and $b2$ before calculating xp.

Another way to look at this is that we are actually saving work – if the two lines are parallel, they cannot intersect, and we therefore know the answer to our question straightaway.

A similar problem arises with vertical lines as shown in Figure 4.5. In this case, the start and the end of the line will have the same X coordinate – however, in calculating the values for $b1$ and $b2$, we need to divide by XE1 – XS1 – again, when these are the same, we will be trying to divide by zero.

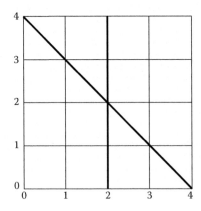

FIGURE 4.5
Line intersection when one line is vertical.

The only way to avoid this is by first testing for the situation of vertical lines. However, the fact that one line is vertical makes the intersection test a little easier because it tells us the X value of the intersection point between the two lines – it must be the same as the X value of the vertical line. This means we already know XP and we can calculate YP from the equation of the sloping line.

At the end of all this, we have an algorithm which is (almost) foolproof:

```
/* Program to test intersection of two line segments
if xs1 == xe1 then
   if xs2 == xe2 then
       print 'Both lines vertical'
       stop
   else                               /* First line vertical
       b2 = (ye2 – ys2) / (xe2 – xs2)
       a2 = ys2  b2*xs2
       xp = xs1
       yp = a2 + b2*xp
else
   if xs2 == xe2 then               /*Second line vertical
       b1 = (ye1 – ys1) / (xe1 – xs1)
       a1 = ys1 – b1*xs1
       xp = xs2
       yp = a1 + b1*xp
   else                          /* Neither line vertical
       b1 = (ye1 – ys1) / (xe1 – xs1)
       b2 = (ye2 – ys2) / (xe2 – xs2)
       a1 = ys1  b1*xs1
       a2 = ys2  b2*xs2
       if b1 == b2 then
          print 'Lines parallel'
          stop
       else
          xp = (a2 – a1) / (b1 – b2)
          yp = a1 + b1*xp
/* Test whether intersection point falls on both lines
if (xs1-xp) * (xp-xe1)  >= 0  and
   (xs2-xp) * (xp-xe2)  >= 0  and
   (ys1-yp) * (yp-ye1)  >= 0  and
   (ys2-yp) * (yp-ye2)  >= 0  then
       print 'Lines cross at' xp,yp
else
       print 'Lines do not cross'
```

Two problems remain. The first is a subtle one. What result will this algorithm produce if the two line segments are exactly the same, or overlap along part of their extent? The answer is that they will probably be judged to be parallel and hence not intersect. Whether or not the lines should be regarded as intersecting or not is actually a matter of definition. The two lines could be regarded as not intersecting, or as intersecting at more than one point.

If we decide that the latter is the correct interpretation, the algorithm will have to be modified to produce this result in this particular circumstance. The second problem is that because of the way the computer stores numbers, the algorithm will occasionally produce incorrect results, especially when the coordinate values are very large or very small. This is explained in more detail in Section 4.4.

4.3 Dealing with Wiggly Lines

Now that we have an algorithm which will detect whether two straight line segments intersect, we can return to our original problem of detecting whether two lines, each made up of a series of segments, intersect. The simple and obvious approach to this problem is to apply our line intersection test to all pairs of segments from all lines. If any pair of segments intersects, this means the lines intersect. This will certainly work, but will take a long time – even with a fast computer. What is more, a lot of these tests will be futile because we will be testing lines which are a long way from each other, and which could not possibly intersect. As an example, consider the three lines in Figure 4.6.

It is clear that lines 1 and 2 cannot possibly cross because they lie in different parts of the map, whereas 2 and 3 clearly do cross. To test whether lines 1 and 2 cross would require us to run our line intersection test 49 times (there are seven segments in each line) so it is worth avoiding this if possible. The way this is done is to perform a much simpler test first to see whether the lines might cross.

In Figure 4.7, a box has been drawn around each line just neatly enclosing it. This structure has various names – minimum enclosing rectangle and

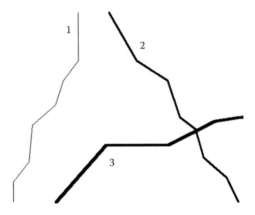

FIGURE 4.6
Testing for the intersection of lines.

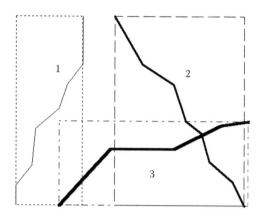

FIGURE 4.7
Minimum bounding rectangle around lines.

minimum bounding box – but in this book, it will be referred to as the minimum bounding rectangle or MBR. The MBRs for lines 1 and 2 do not intersect – this proves that these two lines cannot intersect and that there is no need to do any further testing.

So, how do we construct the MBR and how do we test whether they intersect? The MBR is defined by just four numbers – the minimum and maximum values of the X and Y coordinates of the line, which give the positions of the rectangle corners as shown in Figure 4.8.

So, if we can find these four values, this will define our box. This is very simple – we write a program which reads through the X and Y values for all the points along the line, noting down the largest and smallest it finds. In pseudocode

```
1. xmin = 99999999999
2. ymin = 99999999999
3. xmax = -99999999999
4. ymax = -99999999999
5. for all X and Y
6.     if X < xmin then xmin = X
7.     if X > xmax then xmax = X
8.     if Y < ymin then ymin = Y
9.     if Y > ymax then ymax = Y
```

Bottom left	xmin	ymin
Bottom right	xmax	ymin
Top left	xmin	ymax
Top right	xmax	ymax

FIGURE 4.8
Definition of corner points of the Minimum Bounding rectangle.

The test for whether the MBRs intersect is also simple. Consider the X values – if the smallest X value of one MBR is bigger than the largest X value of the other, then the MBRs cannot intersect. The same sort of logic applies to the other extremes, giving this test

```
1. if xmin1 > xmax2 OR xmax1 < xmin2 OR ymin1 > ymax2 OR
ymax1 < ymin2 then
2.     MBRs CANT INTERSECT
3. else
4.     MBRs DO INTERSECT
```

This test would show that lines 1 and 2 cannot possibly intersect. But what about if the MBRs do intersect – does this mean the lines must intersect, and it is simply a case of finding out where? As line 3 shows that the answer is no. The MBR for this line overlaps those of both lines 1 and 2, although the line itself only intersects with line 2.

So, in the case of lines 2 and 3, our initial test has not helped, and we will still have to run our line intersection. To avoid having to run the full line intersection program for every pair of segments in lines 2 and 3, there are two things we can do. First, we can extend the idea of using the MBR from the whole line to each pair of line segments. If we take the first segments in lines 2 and 3, we can imagine drawing a box around them as we did with the whole line – however, here the corners of the boxes are defined by the end points of the lines, so there is no calculation needed. Our MBR test can therefore be done simply on the end points of the lines.

An alternative is to split the line into what are called monotonic sections – sections in which the X and Y values steadily increase or decrease. Figure 4.9 shows line 1 and a single segment from line 2. As the figure illustrates, it is possible for a segment to intersect a line more than once, and so the segment from line 2 would need to be tested against every segment in line 1.

However, if we divide line 1 into monotonic sections, as shown by the boxes, then within each of these sections, the segment from line 2 can only intersect line 1 once at most. This means that if an intersection is detected, we can move on to consider the next monotonic section.

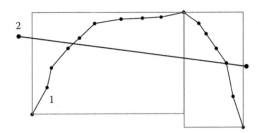

FIGURE 4.9
Monotonic sections of lines.

Later in this chapter, we will present a much better line intersection algorithm than the one which has been described so far. However, the approach which has been described so far is very useful for making clear a number of issues which arise in the design of algorithms for spatial data. The first point about any algorithm is that it must take into account all possible situations, including any special cases. For instance, we saw that a very simple algorithm would fail if one of our lines happened to be vertical. The second point is that an algorithm should be efficient, that is, it should solve the problem as quickly as possible. We may have to run a line intersection algorithm many thousands of times, and even with fast modern computers, this will take a long time if our algorithm is inefficient. There are basically two ways to make an algorithm efficient – to do whatever calculations are necessary as quickly as possible, and to do as few calculations as possible. We will see examples of the first approach in later chapters, but we have already seen that we can improve our line intersection algorithm by including simple checks to identify cases when lines could not possibly intersect, thus saving the need to run the full line intersection program on them. Thus, by doing what at first sight seems to be extra work, such as finding the minimum bounding rectangle of a line, we may in fact save the work overall.

In some situations, it may even be useful to store additional data to speed up processing. Wise (1988) describes a package which allowed users to access the digitised boundaries of the wards of England and Wales. The software started by drawing a map of England and Wales on the screen, and allowing users to select their area of interest by defining a rectangular window, which would then be drawn in greater detail. Once the area of interest had been identified, the boundaries for that area could be written out as a file. This involved a lot of line intersection operations. For instance, it was necessary to decide which ward boundaries fell inside the rectangular window (and had to be drawn on screen), which fell outside (and could be ignored), and which fell across the border (and had to be cut). This meant checking every line in the database against the four lines making up the window. This was a case where the initial test of the minimum bounding rectangles made an enormous difference, and was so important that the MBR for each line in the database was actually stored along with the line itself to save recalculating it each time.

4.4 Calculations on Lines: How Long Is a Piece of String?

We saw in Chapter 3 that we can calculate the length of a line simply by using the Pythagoras equation to calculate the length of each segment and summing these lengths. However, because of the way that computers store numbers, these calculations will sometimes produce the wrong answers. To illustrate, consider the simple example of two points with the following coordinates:

```
Point 1:0,120
Point 2:0,121
```

For simplicity, the X coordinates have been made the same, so we can see that the distance between the points is simply 1 – the difference between the Y coordinates. We can get our computer to check this using a simple program:

```
procedure distance(x1,y1,x2,y2)
d = square_root((x1-x2)*(x1-x2) + (y1-y2)*(y1-y2))
```

This program is very simple to code in any programming language. To illustrate the problems that can arise, a QBASIC version was run on an IBM PC. Putting in our coordinate values as above gave the answer 1 as expected. The same was true with Y values of 1200 and 1201, 12,000 and 12,001 and so on. However, when Y values of 120,000,000 and 120,000,001 were used, the answer was given as 0. Values of 12,000,000 and 12,000,001 gave 1, but the same two numbers divided by 10 (1,200,000 and 1,200,000.1) gave 0.125 instead of 0.1.

So what is happening? The problem is not with our formula, but with the way the computer stores numbers.

We saw in Chapter 1 that the computer stores numbers using binary digits, or bits, in fixed length units of 8 (a byte), 16, 32 or 64 (words of varying length). The largest whole number which can be stored in n bits can be calculated using the formula

$$2^n - 1$$

The most usual word length is 32 bits, which can store a number up to 4,294,967,294 (just over 4 billion). If we want to be able to store negative numbers too, then one of the bits in the word is reserved for this purpose. This is normally the leftmost bit, and if this is 0, the number is positive; if it is 1, the number is negative.

This is fine for whole numbers (or integers). But, what about numbers with a decimal point in (sometimes called real numbers)? Let us turn to the familiar world of base 10 numbers first for an explanation. With a number like 24.5, we have to store the digits 2, 4 and 5 and also the fact that there is a decimal point after the first two digits. One way of doing this is to move the decimal point so that it is to the left of the first digit, and then record how many places it has been moved. So 24.5 can be stored as

$$24.5 = 0.245 \times 10^2$$

This system is widely used in the scientific literature because it is a convenient way of dealing with very large or very small numbers. For example, the mass of an electron at rest can be represented as $0.9109534 \times 10^{-30}$ kg, which is far easier to deal with than a string of 30 zeroes followed by the digits

9109534. This is the method of storing all real numbers on the computer, regardless of how large or small they are, but using binary rather than decimal arithmetic. This method of storage real numbers is called floating point because the point 'floats' to the front of the number and the term 'floating point' is often used to refer to numbers which are not integers.

To explain the problems which can arise with floating point, we will stay with our decimal example for the moment. To store a number in floating point, we actually have to store two numbers – the actual digits themselves (called the mantissa) and the number of 10s they must be multiplied by (called the exponent). In the world of pen and paper, there is no limit to the length of either the mantissa or the exponent. However, in the world of the computer, we are limited to what we can store in our word. This is like having 32 boxes, each of which can hold one digit. Some of the boxes will be used for the mantissa and some for the exponent. If we split them evenly, we can store numbers with 10,000,000,000,000,000 digits in them (the maximum exponent size), but in the mantissa there is only room to store the first 16 of them. It is more usual therefore to reserve more spaces for the mantissa than the exponent. Even with a 2/30 split, we can store numbers with 100 digits, but only record the first 30.

Exactly the same principle applies in the binary system, but in this case, each box can only hold one bit rather than a decimal digit. The same problem also arises – this technique allows us to store very large or very small numbers, but only the first few digits of those numbers can be recorded – the rest are simply lost. So, when we attempt to calculate distances using coordinates like 1200 and 1201, we are safe, but as soon as the numbers get up to the range of 12,000,000 and 12,000,001 the results become unreliable.

These problems are compounded whenever calculations are performed – the loss of digits in the numbers means the answer is slightly wrong, and if this is then used in another calculation, that answer will be even more wrong and so on. It is not just calculations such as distance which are affected. The line intersection algorithm described in Section 4.2 was based on a series of calculations, any of which could be affected by computer precision problems. For example, the final part of the code for line intersection is the final set of tests to see whether the computed intersection point lies on both the line segments:

```
if (xs1-xp)*(xp-xe1) >= 0 and
   (xs2-xp)*(xp-xe2) >= 0 and
   (ys1-yp)*(yp-ye1) >= 0 and
   (ys2-yp)*(yp-ye2) >= 0 then
        print 'Lines cross at' xp,yp
else
        print 'Lines do not cross'
```

Let us consider the first line, which tests whether the X coordinate of the intersection point (xp) lies between the X coordinates of the two ends of the first line segment (xs1, xe1). xp has been calculated earlier in the algorithm.

If the coordinate values of the line segments contain more than seven significant digits, and their values are very similar, then xp may appear to be the same as either xs1 or xe1, when in fact it is not. This will make the result of the overall calculation 0, making it appear that the intersection point falls on this line segment, when it may actually fall just outside it. Conversely, xp may actually be exactly the same as either xs1 or xe1, but because of the lack of precision in the calculations, it appears to be slightly different. Again, this could lead to the wrong conclusion being reached in the test.

These problems can easily arise in real-world applications. Global mapping applications such as Google World and Bing Maps record locations using decimal degrees:

```
<coordinates >
   -122.0848938459612,37.42257124044786,17
</coordinates >
```

The circumference of the earth is approximately 49,075 km at the Equator so 1° represents approximately 111 km. This means that 1 m is approximately 0.00009°, so using single precision would mean that points closer together than a metre might appear to be in the same place. Another example is the British National Grid. This has its origin somewhere near the Scilly Isles and records positions in metres, so that by the time you reach Scotland, the Y coordinates of points will be around 800,000 – in the Shetlands they exceed 1,000,000 m, which again is on the limit of standard computer precision this time by being too large rather than too small.

One solution is to use a local origin for the coordinates of points. In the example of the Shetland Islands, all Y coordinates could have 1,000,000 subtracted from them, so that the important variations between positions within the area fall within the computer precision. An alternative solution is to use what is called double precision, if this is supported by the GIS package. This uses a longer word for each number, not only increasing the number of significant digits which can be stored but also doubling the amount of storage space required. Judging by the 13 digits used to report positions from web mapping packages, this appears to be the solution adopted in these cases. GIS designers can also help avoid these problems by designing algorithms which explicitly deal with issues of precision. This may involve replacing all the simple tests for equality (==) and inequality (>, >= etc.) with more complex tests which include tolerance values rather than testing for absolute equality or inequality.

4.5 Line Intersection: How It Is Really Done

The simple algorithm for line intersection was introduced in Section 4.1 because it is a good way to illustrate the difficulty of producing an algorithm for what appears to be a simple problem. However, it is not actually a very good algorithm because it depends on calculating the intersection point between line segments in order to determine whether the segments actually intersect. If the lines do not intersect, then working out where they might have done is wasted effort and so if we could find a way of detecting whether the lines intersect first this would be faster. The algorithm also makes use of division, which is often much slower to carry out in the CPU than other operations such as addition or multiplication. Having fewer calculations will not only increase speed but also reduce the problems caused by fixed precision, which were described in the previous section because these errors tend to accumulate as we do more operations. For example, assume we wish to add 2,300,000 and 2. As in Section 4.4, the explanation will ignore the fact that computers work in base 2 and use base 10 for simplicity. We saw in Section 4.4 that in floating point format 2,300,000 would be represented as 2.30×10^7. This is often written as 2.30e7, where e is short for exponent. To perform the addition, the processor will make the exponent value the same for both numbers:

```
2.30e7 + 2.0e0 = 2.3e7 + 0.0000002e7
```

Owing to the limited precision, the result of this might well be 2.30e7. This means there is an error of 2 in the answer. If this answer is then added to another number, then this error will carry over into the result. In addition, the second addition may add more error. If the result is multiplied by another number, then the error is also multiplied and therefore becomes larger.

Rather than calculating the intersection point of two lines, let us see if we can identify what conditions must be satisfied for lines to intersect. Effectively, there is just one:

1. The end points of each line will be on opposite sides of the other line.

To see why this is sufficient, consider Figure 4.10. Start by considering lines A and B. Line B has its end points on either side of line A. However, both of the

FIGURE 4.10
An illustration of the conditions required for line segment intersection.

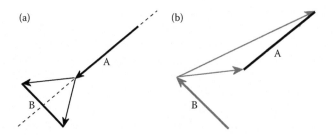

FIGURE 4.11

Test for whether two points are on different sides of a line. In (a), B straddles A and the vectors from A to the end points of B turn in opposite directions. In (b), A does not straddle B and so the vectors from B to the end points of A both turn in an anticlockwise direction.

ends of line B lie beyond line A and so the two lines do not intersect. If you now imagine line B being moved along as shown by the arrows, the two lines can only intersect when the end points of A also straddle B as shown in position B′. If B moves past the end of A, as at B″, then the two lines no longer intersect. So, if we have a test for whether two points are on different sides of a line, we will have a test for line intersection. One such test is illustrated in Figure 4.11.

Figure 4.11a shows how we can test whether segment B straddles segment A. Instead of being considered simply as a line, segment A is now treated as a vector, which is a line which has a direction, as shown by the arrow. Vectors are then drawn from the end of A to each end of line B. If you imagine yourself travelling along line A, then to get to one end of line B you would turn to the left, and to get to the other end you would turn right because segment B straddles the infinite line running through line A. In contrast, if we consider segment B as a vector and connect it to each end of line A (Figure 4.11b) we can see that in both cases, we turn to the right, because both ends of line A lie to the right of line B. If A had been to the left of B, both lines would have turned to the left.

Now, we have a way of testing whether two lines intersect or not, which is as follows:

- Consider line 1 as a vector and construct vectors to each end of line 2.
- If these turn in opposite directions from each other the ends of line 2 straddles line 1.
- Consider line 2 as a vector and construct vectors to each end of line 1.
- If these turn in opposite directions from each other the ends of line 1 straddles line 2.
- If – and only if – both lines straddle each other, the lines intersect.

Now, all we need is a way of testing which way one vector turns relative to another vector. This would seem to be a complicated problem, requiring the calculation of angles which is both time-consuming and prone to small errors. However, considering the segments as vectors is the key because vector algebra provides a very elegant and simple solution to this problem, in the form of what is called a cross-product. The calculation will be illustrated with reference to a simple example as shown in Figure 4.12a, which shows three points P0, P1 and P2.

Point P0 is the origin of two vectors that connects this point to points P1 and P2, respectively. If the locations of the points are $(x0,y0)$, $(x1,y1)$ and $(x2,y2)$, respectively, then the cross-product of the two vectors is as follows:

$$cp1 = (x1 - x0) \cdot (y2 - y0) - (x2 - x0) \cdot (y1 - y0)$$

To understand how this calculation can help, Figure 4.12b shows a simplified version of this diagram in which point P0 is the origin, which means that $x0$ and $y0$ in the above equation are zero and so the equation simplifies to

$$cp1 = x1 \cdot y2 - x2 \cdot y1$$

The vectors are then rotated until they run along the axes (Figure 4.12c). Both $x2$ and $y1$ become zero and the equation becomes

$$cp1 = x1 \cdot y2$$

This is a positive value and this can only occur if the second vector (defined by $x2,y2$) has a direction which is anticlockwise from the first. If we change the first vector so that it runs to the left from the origin, $x1$ will have a negative value and this will make the cross-product negative. To see how we can apply this to our line intersection test, consider Figure 4.13, which is a redrawn part of Figure 4.11. This time, however, the four points which define the two lines have been numbered.

We wish to know whether the vector from the end of segment B (point 2) to one end of segment A (point 3) turns right or left. To find out, we can

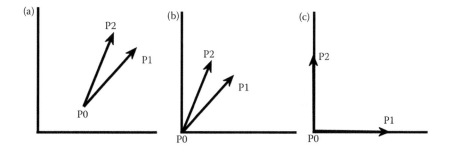

FIGURE 4.12
Illustration of cross-product calculations. The principle can be more easily understood if the two lines (a) are shifted to the origin (b) and then rotated to be parallel with the axes. This simplifies the cross-product expression as explained in the text.

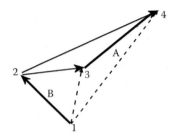

FIGURE 4.13
Testing for the intersection of lines A and B using cross-products.

compare the direction of the lines 1–2 with that of lines 1–3 using our cross-product:

$$cp1 = (x2 - x1) \cdot (y3 - y1) - (x3 - x1) \cdot (y2 - y1)$$

This will return a negative result because the line to point 3 is clockwise from the line to point 2. If we repeat this test with points 1, 2 and 4, we will also get a negative result because the line from 1 to 4 is clockwise with respect to the line from 1 to 2. Our line intersection algorithm can now be summarised as follows:

```
Given two lines with end points P1, P2 and P3, P4
D1 = crossproduct(P3 to P4 to P1)
D2 = crossproduct(P3 to P4 to P2)
D3 = crossproduct(P1 to P2 to P3)
D4 = crossproduct(P1 to P2 to P4)
IF ((D1 > 0 AND D2 < 0)) OR (D1 < 0 AND D2 > 0)) AND
    Intersect = TRUE
ELSE
    Intersect = FALSE
```

As with the algorithms described earlier, this algorithm would need to consider special cases such as when lines touch. However, these are not considered here and those interested may follow this up via the material in the Further Reading section. We finish this chapter by considering how this algorithm is used to test whether two segments intersect as part of a widely used algorithm for finding all the intersections between a number of segments such as those shown in Figure 4.14.

The extent of each segment along the X axis is shown by the vertical lines through its end points. As with the MBR, these extents tell us which lines may be candidates for intersection because only lines whose extents overlap can have any possibility of intersecting. It is easy to see that line A cannot intersect with any other line because its extent does not overlap with any of the others. Lines B and C have overlapping extents, but do not intersect whereas C and D do. The brute force approach to testing for intersections would compare each

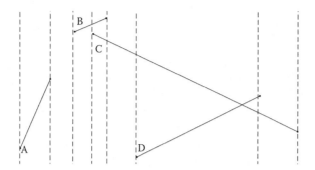

FIGURE 4.14
The use of the extent of lines in the X direction to identify candidates for intersection.

line with all the others to test for possible overlaps between extents. However, by just considering the extents in one direction, we can use a method known as a plane sweep algorithm which is much more efficient. In such an algorithm, you have to visualise a vertical line sweeping across the lines from left to right. A segment is not considered as a candidate until the sweep line has reached it and as soon as the sweep line passes its end, it ceases to be considered as a candidate. In the case of Figure 4.14, the algorithm would add line A to the list as it reached its start point and drop it as it reached its end. In between these two positions, the sweep line would not encounter any other lines and so line A would never be considered as a candidate for intersection. When the sweep line reaches the start of line C, it will have B on its list and so it will have to start testing lines for possible intersection.

One of the best known of the plane sweep algorithms is the Bentley–Ottmann algorithm for line intersection. To see how it operates, let us add a few extra lines to produce the set in Figure 4.15.

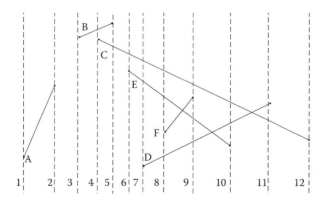

FIGURE 4.15
Start and end of each segment identified as events for the running of the Bentley–Ottmann line intersection algorithm.

The lines passing through the end points of each segment have now been numbered. The algorithm has been described as a line sweeping from left to right, but this is simply a way of visualizing how it works. In practice, the algorithm only ever needs to do anything at the start and the end of a segment and when two segments intersect. It therefore begins with a list of the starts and ends of all the segments sorted by their position on the X axis, which we will call the event list. It will then consider each event in turn. When the event is the start of a segment, the segment is added to a second list, which we will call the active segment list. When the event is the end of a segment, this segment is taken off the active event list. At any given time, the active segment list contains all those segments whose x-extents overlap and which might therefore intersect. To understand a key element of this algorithm, consider event 8 in Figure 4.15. At this point, there are four segments in the active segment list – C, D, E and F. However, the algorithm does not need to test every segment against every other one. To understand the reason, consider Figure 4.16 which shows the part of Figure 4.15 just before lines E and F intersect

At the point just before they intersect, these two lines are very close together in the Y direction. This means that if we have a list of the lines sorted in terms of their position on the Y coordinate, only lines which are neighbours on this list need to be considered as candidates for intersection. To see why this will always work, imagine what would happen if we insert another line in between E and F as shown. One of two things will happen to this line.

1. It will terminate before the point at which E and F cross, in which case, E and F become neighbours again.
2. It will cross either E or F. Once it has done this, E and F become neighbours again, and the new line cannot intersect either of them again until they themselves have intersected.

Strictly speaking, there is a third possibility, which is that the new line intersects both E and F at the point at which they cross. However, this is one of those special cases that a real program would have to deal with, but which we can ignore for the sake of simplicity.

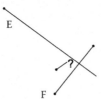

FIGURE 4.16
Illustration of why only neighbours on the sorted active list need be considered as candidates for intersection.

This means that if we keep track of the order of the segments on the active segment list in terms of their position in Y, we only need to consider immediate neighbours on this list as candidates for intersection. The difference this makes can be considerable. If we have n segments on the list, then considering all pairs means $(n^2 - n)/2$ comparisons. However, if we only consider neighbours, then we have only $(n - 1)$ comparisons to make. For instance, when there are 20 segments in the list, there are 190 possible pairs, but only 19 of these involve neighbours.

We now have enough background to describe how the Bentley–Ottmann plane sweep algorithm works in full. The algorithm will begin with two lists as follows:

```
Event List: {1,2,3,4,5,6,7,8,9,10,11,12}
Active Segment List: {}
```

The first event in the list is the start of segment A, so this will be added to the event list, but then removed again at event 2. Event 3 is the start of segment B, so B is added to the active list. Event 4 is the start of segment C, which also has to be added to the active list. However, this list now contains another segment and so we must decide whether C is above or below B. The only way to do this is to find the point at which B intersects the sweep line and see whether this is above or below the Y coordinate of C. Note that we cannot use the Y coordinate of the start of B for this because as Figure 4.14 shows this was below the start of C, whereas line B is now above C. We are going to have to do this calculation for each segment on the active list every time we encounter the start of a segment, so it might seem that by having all these extra intersection tests to do we are losing all the advantages of using a plane sweep. However, this is a very simple intersection calculation to do for two reasons:

- We know that segment B and the plane sweep must intersect by definition so we do not have to test for this.
- We only need the Y value of the intersection because we already know the X value since it is the X coordinate of the current event.

In fact, the value can be determined with a single calculation. If the coordinates of segment B are $X0,Y0$ for the start and $X1,Y1$ for the end, and the event (i.e., the start of segment C) has an X coordinate of XE, the Y value we want is given by

$$Y = Y0 + \frac{(XE - X0)}{(X1 - X0)} \cdot (Y1 - Y0)$$

In this case C will be added below B in the list. The next step of the algorithm is to test whether the newly inserted segment intersects with its neighbours.

There is only one neighbour, so B and C are tested for intersection. They do not intersect, so the algorithm proceeds to event 5, the end of B.

The next three events will add segments C, E and D to the active segment list (ASL), which at event 7 will look like this

$$ASL = \left\{ \begin{matrix} C \\ E \\ D \end{matrix} \right\}$$

C and E are tested but do not intersect, but when E and D are tested we have our first pair of segments which intersect. To see what needs to happen now, look at the point at which E and D intersect. At this point, they cross and therefore will need to swap places on the ASL.

$$ASL = \left\{ \begin{matrix} C \\ D \\ E \end{matrix} \right\}$$

This swap will make D a neighbour of C for the first time and as Figure 4.14 shows these segments do indeed intersect. However, we cannot swap C and D on the ASL yet because when we add the start of segment F to the ASL at event 8, the positions of C, D and E on the sweep line will be recalculated and C and D will be swapped back. The only way to ensure that the swap occurs at the correct time is to add the intersection between E and D as a new event on the event list and mark it as being an intersection event in between current events 9 and 10. When this point is reached, the swap is made and both segments are tested against their new neighbours at which point the intersection between C and D is detected.

This brief overview of the Bentley–Ottmann algorithm shows how important it is in the design of good algorithms that as little work as possible is done to achieve the correct results. In this case, this was achieved by using a sweep line to process the segments in sequence, and also by ordering the segments under active consideration according to their relative Y coordinates. This algorithm has also introduced a new development in the role of data structures. So far, data structures have been considered as a means of storing data, which is largely unchanged, such as the coordinates of the line segments shown in Figure 4.14. However, with the event list and the active segment list, we have two examples of structures that are storing data, which is constantly changing as points are added, deleted and swapped around. In fact, a simple list such as what we used for the XY coordinates of line data is a very poor way of storing such data because every addition, deletion or swap means moving large parts of the list around. We need a far more flexible way of storing data in this situation and this will be described in Chapter 11 in the context of analyzing lines which form part of a network.

Further Reading

The discussion of the line intersection problem in this chapter is largely based on the material from Unit 32 of the NCGIA Core Curriculum, and on the studies of Douglas (1974) and Saalfeld (1987). Bentley and Ottmann (1979) is the original source paper for the algorithm which is named after them, but De Berg et al. (1997) have a good description of it in their Chapter 2. Park and Shin (2002) discuss how the fact that lines in GIS are formed of connected segments poses particular issues for the line intersection problem but also provides some ways to speed up the Bentley–Ottmann approach. Nievergelt (1997) provides an excellent introduction to the development of computational geometry, and uses the example of finding the nearest neighbour among a set of points to illustrate some of the features of designing algorithms for handling spatial data. The problems caused by limited computer precision are well illustrated by the example discussed by Blakemore (1984) in the context of locating points in polygons. Greene and Yao (1986) have proposed a solution to these problems, by explicitly recognising that vector data have a finite resolution and that operations such as line intersection must take this into account. Worboys (1995) provides a brief summary of their approach. Schirra (1997) points out some of the weaknesses of standard solutions and the need to include tolerance values in tests for equality. The paper by Wise (1988), while mostly about a particular piece of computer hardware, also includes a discussion of the use of the minimum bounding rectangle in designing a system which allowed users to select windows from a national data set of area boundaries.

5

Vector Algorithms for Areas

5.1 Calculations on Areas: Single Polygons

We saw in Chapter 3 that area features are stored in GIS as polygons, represented using a single closed line, or a set of connected links. In this chapter, we are going to consider some of the fundamental operations involving areas in GIS, starting with finding out how large an area is.

For the moment, we will assume that the polygon is stored in the simplest possible form – as a line which closes on itself like the example in Figure 5.1.

This line is stored as a series of five points which form the ends of the five straight segments. Three of these form the upper part of the boundary, and two the lower half. If we calculate the area under the 'upper' segments, and subtract the area under the 'lower' segments, we will be left with the area of the polygon itself, as shown diagrammatically in Figure 5.2.

This is fairly easy for us to understand visually, but we must remember that the computer does not 'see' the area in the way we do – all it has is a series of X,Y coordinates for each of the five points on the polygon boundary. I will describe how we can determine which parts of the boundary are above and which below the figure in a moment – first, we need to know how to calculate the area under a line segment, so it is back to the maths classroom!!

Let us start with the first line segment, shown in Figure 5.3.

The area under this line can be divided into two parts, as shown by the lines. The lower part is a rectangle, with width $(x2 - x1)$ and height $y1$, so its area is simply

$$(x2 - x1) \cdot y1$$

The upper part is a triangle, whose area is half the area of the small rectangle as shown (Figure 5.3). The width of this rectangle is $(x2 - x1)$ as with the first one. Its height is $(y2 - y1)$ and so half its area will be

$$(x2 - x1) \cdot \frac{(y2 - y1)}{2}$$

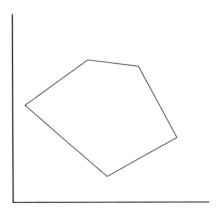

FIGURE 5.1
Single polygon.

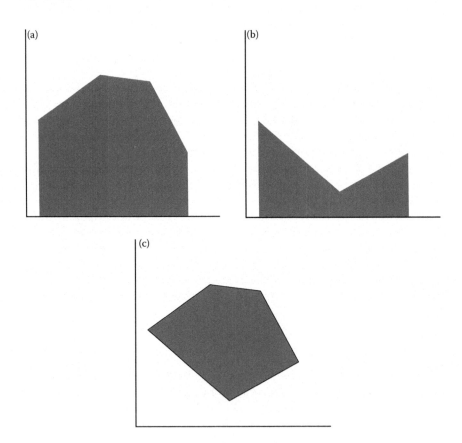

FIGURE 5.2
Calculating the area of a polygon. Area beneath lower segments (b) is subtracted from area under upper segments (a) to calculate polygon area (c).

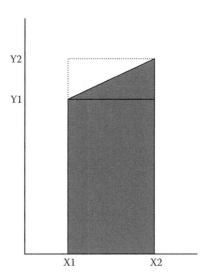

FIGURE 5.3
Calculating the area under a single line segment.

If we add these two areas together, we get the following formula:

$$(x2 - x1)\left[y1 + \frac{(y2 - y1)}{2} \right]$$

which can be simplified to the following:

$$(x2 - x1) \cdot \frac{(y1 + y2)}{2}$$

We can apply the same formula to each line segment replacing $x1,y1$ and $x2,y2$ with the appropriate values for the start and end points of the segment. So, how do we know which segments are below the polygon, and need to have their area subtracted from the total? To answer this, let us simply apply our formula to the segment between points 4 and 5 (Figure 5.4) and see what happens. In this case, the formula for the area will be

$$(x5 - x4) \cdot \frac{(y5 + y4)}{2}$$

The first term will give a negative value because $x5$ is smaller than $x4$ and so the area under this line segment will have a negative value. In other words, by simply applying the same formula to all the line segments, those under the polygon will have a negative area while those above will have a positive value. This means that we do not have to take any special action to identify which segments are above and which below the polygon – we simply apply

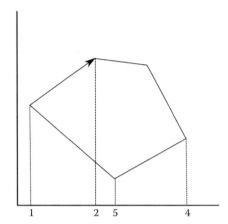

FIGURE 5.4
Area under line segments which are 'above' and 'below' the polygon.

our formula to each segment in turn and by summing them as we go we will arrive at the area of the polygon.

This simple algorithm will work as long as the polygon boundary is stored as a sequence of points in a clockwise direction – if the points are stored in an anticlockwise direction, the answer will be numerically correct, but be negative rather than positive!

Of course not all polygons have such simple boundaries as this one. However, as long as the boundary is stored as a clockwise sequence of points, the algorithm will still work correctly no matter how convoluted or strangely shaped the boundary.

5.2 Calculations on Areas: Multiple Polygons

In many GIS systems, area features are not stored using simple polygons, but using the link and node structure described in Chapter 3. This data structure is particularly useful when a GIS layer contains a series of polygons which border each other because it is only necessary to store the shared boundaries between polygons once, as shown in Figure 5.5.

This shows two polygons, A and B, whose boundaries are made up of links (numbered 1, 2 and 3) which meet at nodes. Each link is stored as a series of points (for which we have an X and Y coordinate), which define a series of straight line segments. In addition, we also know which node the line starts and ends at, and which polygon is on its left and right.

As we saw in the previous section, if the boundary of a polygon is a single line running in a clockwise direction, the area of the polygon can be calculated by working out the area under each line segment using the following formula:

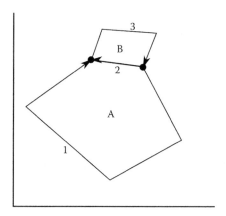

FIGURE 5.5
Storage of two neighbouring polygons using the link and node data structure. The arrows show the direction of digitising of each link.

$$(x2 - x1) \cdot \frac{(y1 + y2)}{2}$$

Here, $x1,y1$ are the coordinates of the start of the line and $x2,y2$ are the coordinates of the end of the line. These areas will be positive for lines 'above' the polygon and negative for lines 'below' the polygon, so by summing them the area of the polygon itself is calculated.

In Figure 5.5, instead of a single line, we have a set of links forming the boundary. However, this does not really affect our formula, which considers one line segment at a time – as long as we can connect our links into a clockwise boundary, the method will work as before.

If we look at polygon B, we will see that its boundary is made up of links 2 and 3, which do indeed go clockwise around the polygon. In this case, we can simply apply the algorithm to the two and we will get the area of polygon B.

However, this is not true for polygon A because link 2 is going in the wrong direction. We can see this from the diagram, but of course all the computers have to go on as per the information we have stored about the links. We thus have two problems to solve – how can the computer tell whether a link is going the right way around a polygon, and if it is not, what can be done about it?

The answer to the first lies in the topological data stored for each link. If a line circles an area in a clockwise direction, then the area will be on the right-hand side of the line – the same logic also applies to the links in a link and node structure. In fact, you may remember we used the same rule when checking the accuracy of street data in the DIME system. Applying this rule, we will be able to detect that link 2 is the wrong way round for calculating the area of polygon A since A is on the left of this link. For calculating the area of B the link is in the right direction, of course.

One option for dealing with this is to reverse the direction of link 2 when doing the calculations for polygon A. However, if we use this approach, we will apply our formula to link 2 twice – once while calculating the area of polygon A, and once while calculating the area of polygon B. The only difference is that link 2 is above polygon A but below polygon B – in the first case, the area under the link must be added to our sum; in the second case, subtracted. A better approach is to calculate the area under the link once, and then use the topological information to decide whether to add this or subtract it from the area of each polygon.

Let us assume that the coordinates of the line as shown in Figure 5.5 are as follows:

	X	Y
Start	4	1
End	3	2

When we enter these figures into the formula

$$(3 - 4) \cdot \frac{(1 + 2)}{2}$$

we get the answer –1.5. The result is negative because the start point is to the right of the end point and so has a larger X coordinate. Link 2 is pointing the right way to form part of a clockwise boundary around polygon B, and the computer knows this because B is on the right of link 2. The same logic applies in this case as with the simple polygon described in Section 5.1 – the negative area ensures that the area under link 2 will be subtracted from the running total. When we try and calculate the area of polygon A, we need to know the area under link 2. This will still be 1.5, but now it must be added to the sum, and not subtracted. Since we know it is running anticlockwise around A (because A is on its left), all we need to do is take the result of the formula and reverse the sign.

This example has used just one link. The general rule is as follows. Take each link in turn and apply the formula to calculate the area under that link. For the polygon on the right, the link is running clockwise and so the area can be used exactly as calculated (whether this is negative or positive). For the polygon on the left, the link is running anticlockwise, so the sign of the area is changed (from negative to positive or from positive to negative) and then the answer added to the sum for that polygon. If you want to check that this works, work out what would have happened if link 2 had been digitised in the other direction – the area should still be negative when link 2 is considered as part of polygon B, but positive when it is part of polygon A.

Some links will only border one 'real' polygon – on the other side of the line will be the 'outside' polygon. An interesting side effect of this algorithm is that if the same rule is followed for this outside polygon as for all the others, it receives an area which is the total of the areas of all the other polygons, but with a negative value.

5.3 Point in Polygon: Simple Algorithm

Now, we will turn to another of those problems which seems simple, but which is difficult to program a computer to solve – determining whether a point lies inside or outside a polygon. A typical example of the use of this algorithm is the matching of address-based data (assuming we can determine a location for the address) with area data such as data from a population census. Typical examples will try and match several points to several polygons, but for simplicity, we will begin with one point and one polygon, as shown in Figure 5.6.

As with the line intersection problem we considered earlier in this series, it is very simple for us to see that only the leftmost point lies within the polygon. However, all the computer can 'see' is the position of the points (as *X,Y* coordinates) and the location of the points defining the polygon boundary.

A number of methods have been developed to test whether a given point lies inside a polygon. In the one which is most commonly used, an imaginary line is constructed extending out in any direction from the point – if this line intersects the boundary of the polygon an odd number of times, the point is inside the polygon. In the example above, the line from the first point intersects the boundary once and so this point is inside the polygon. The line from the second point intersects twice and so this point is outside. The line from the third point does not intersect at all. Mathematically speaking, zero can be

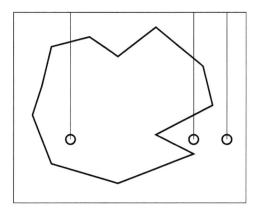

FIGURE 5.6
Point in polygon test.

considered an even number because there is no remainder when it is divided by 2, and so the test works in this case too. In simple terms, the algorithm is

```
1. n = 0
2. for each line segment
3.    test intersection with 'ray' from point
4.    if lines cross, n = n + 1
5. if n is odd, point is in polygon
```

Each line segment in turn is tested for intersection with the 'ray' from the point and a variable called n is used to count these intersections. An interesting detail of this algorithm is the method used to determine whether the number of intersections indicates that the point is inside or outside. The method above counts the intersections but then must determine whether the total is an odd or even number. It is also possible to write the algorithm to give the answer directly by using what is sometimes called a 'toggle' – a variable which switches between two possible values:

```
1. n = -1
2. for each line segment
3.    test intersection with 'ray' from point
4.    if lines cross, n = n * (-1)
5. if n == 1, point is in polygon
```

In this example, variable n is set to –1. If an intersection is found, n is multiplied by –1 changing its value to +1. If a second intersection is found, n is again multiplied by –1, changing its value back to –1. The value of n therefore toggles between the values of –1 and +1 each time an intersection is found – an odd number of intersections will leave the value at +1.

This sort of detail will not affect the speed of the point in the polygon test very much but the way in which the line intersection test is done will. We saw in Chapter 3 how to test whether two lines intersect and so it would seem simple enough to use this program to test each segment of the boundary line in turn against our ray. This would certainly work, but as with the line intersection test itself, there are a number of things we can do to produce a much faster method.

The first of these is to use the MBR test (Figure 5.7).

As shown above, we can begin by testing the points against the smallest rectangle which encloses the polygon – a point which falls outside this rectangle cannot lie inside the polygon and there is no need to perform any further tests. As this example shows, it is possible for a point to lie inside the MBR, but outside the polygon itself. However, if we are testing a single point against a large number of polygons (as with the example of finding which census area an address lies in), this initial test will help us reject a large number of the polygons very quickly. Once we have our candidate polygons, there are other things we can do to improve the speed of the line intersection test.

The standard line intersection method first determines the equations defining the two lines, calculates the point at which infinite lines defined by these

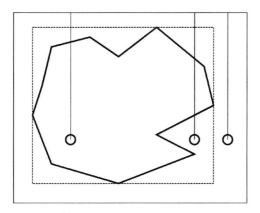

FIGURE 5.7
Use of the minimum bounding rectangle for the point in polygon test.

equations intersect, and then tests whether this point lies on both the actual, finite lines being tested. However, in this case, we can determine the characteristics of one of the lines in question – the ray extending from the point – and as Figure 5.6 shows, a vertical line is used. This alone would speed up the standard line intersection test, since we know the X value of the intersection point and would only have to work out the Y value.

We can make further improvements by developing a special algorithm for line intersection in this particular case. Since the ray is vertical, a line segment can only intersect with it if the ends of the segment lie on either side of the ray. If we denote the X value of the ray as xp (since it is the X coordinate of our point), intersection is only possible if xp lies between $x1$ and $x2$ and one way to test for this is as follows:

```
((x1 < xp) and (x2 > xp)) or ((x1 > xp) and (x2 < xp))
```

Figure 5.8 shows what happens when we apply this test to the leftmost point in Figure 5.6. Two line segments have passed the test because both intersect the infinite vertical line which passes through the starting point of the ray. We are only interested in those segments which intersect the finite vertical line which passes upwards from the starting point. To find out which these are, we will need to perform the full line intersection calculations, and then test whether the intersection point is above or below the starting point of the ray. The advantage of our initial test is that we will only have to do this full set of calculations for a small proportion of the segments making up the polygon boundary.

In making this test, we have to be careful about the special case where the ray passes exactly through one end of the line, as shown in Figure 5.9.

In this case, Xp is equal XB and if we perform the test as shown above, neither line will pass. The test must therefore be modified to become

```
(x1 = xp) or ((x1 < xp) and (x2 > xp)) or ((x1 > xp) and (x2 < xp))
```

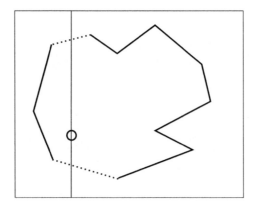

FIGURE 5.8
Line segments selected by initial test on *X* coordinates.

Note that we only test one end of each segment to see whether it is the same as *Xp*. In the above example, the first segment will fail the above test because *X*1 in this case will be the value for *XA* – however, for the second segment, *X*1 will be equal to *XB* and so the segment will pass the test. If we tested both ends of the segment for equality with *Xp*, both segments would pass the test and we would still miscount.

In fact, the test as described above will run into problems because of the limited precision of floating point operations, as described in Chapter 3. In particular, testing for exact equality ($x1 = xp$) will be particularly sensitive, and it is far safer to test for $x1$ being within a very small distance of xp. To do this, we replace the term in the first brackets with

```
(x1 > (xp – d) and x1 < (xp + d))
```

where *d* is a very small distance.

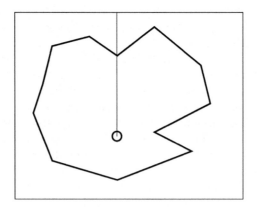

FIGURE 5.9
Special case of a point lying directly beneath a vertex on the polygon boundary.

These alterations to the basic original algorithm will not make much difference if we have one point and one polygon, but in more typical uses of the algorithm, there will be several points and several polygons. For example, in a study of disease incidence in Sheffield, the staff at Sheffield University used a GIS to find which of the 1100 census EDs (Enumeration Districts) each of 300 patients lived in. This meant testing each of 300 points against 1100 polygons – on average, a match will be found for each point halfway through the set of polygons, but this still gives 165,000 tests to be performed. However, if we apply our simple MBR test, then in most cases, only three or four polygons will be possible candidates, meaning we only have to run the full test about 1200 times. The saving in time will more than compensate for the extra work of doing the MBR tests.

5.4 ... and Back to Topology for a Better Algorithm

So far, we have only considered how to decide whether a single point lies inside a given polygon or not. However, rather than a single polygon, it is more usual to have a GIS layer containing numerous polygons, stored using the link and node method described in Chapter 3. What is needed is to find which of the many polygons a point falls into (Figure 5.10).

One approach would be to test the point against each polygon in turn. We would have to construct the boundary of the polygon, by retrieving all the

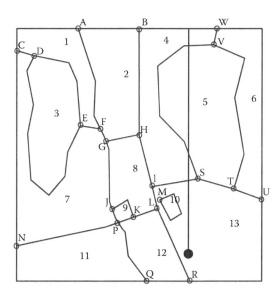

FIGURE 5.10
Point in polygon test for multiple polygons.

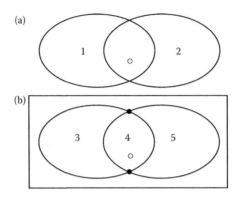

FIGURE 5.11
The consequence of the planar enforcement rule. (a) Without planar enforcement the point lies within both polygon 1 and polygon 2. (b) With planar enforcement the point lies within polygon 4.

relevant links, find the MBR and then if the point fell inside this, carry out the full point in polygon test.

The problem with this approach is that most links would be processed twice. Figure 5.10 shows a point in polygon 13. The MBR test would indicate that polygons 12, 13, 4, 5 and 6 are all potential candidates for the polygon which the point falls in. All these polygons neighbour one another and so many of the links which make up their boundaries would be tested against the ray twice. For example, the link between I and S would be tested as part of the boundary of polygon 13 and again as part of polygon 5. So, is there a way of avoiding this double processing? There is and again the key is in the use of topology. We have already seen that polygon boundaries can be stored using links which join at nodes, an idea developed by applying Poincare's model of maps as topological graphs. However, this model also means that whenever lines cross, they *must* have a node at the crossing point. To explain what this means, and why it is important, consider Figure 5.11.

Figure 5.11a contains two polygons which overlap but which do not have nodes where their boundaries cross. The problem with this is that the rules which we used for checking the accuracy of a digitised map no longer apply. For example, in checking the DIME file, it could be assumed that each line formed part of the border between two areas, but in the upper diagram, the same line can sometimes form the border between polygon 2 and the inside of polygon 1, and also between polygon 2 and the 'outside world'. For the simple topological rules to work, GIS systems use what is called 'planar enforcement' – this means that lines are all considered to lie in the same plane, and so if they cross, they must intersect – one line cannot go over another one.

This is illustrated in the Figure 5.11b, where nodes have been placed where the lines cross. This actually creates *two* new polygons and modifies the two original ones. Polygons 1 and 2 in the upper figure are each split into two parts, and the area in common between them becomes a polygon in its own right (polygon 4 in the diagram). In addition, since the rules state that each link must lie between two polygons; there has to be an 'outside' polygon.

The difference between the two is very relevant to the point in polygon problem. Consider the point which is drawn on Figure 5.11. In Figure 5.11a, this lies in both polygon 1 and 2 – in Figure 5.11b, it is only in polygon 2. In fact, if we enforce the rule of planar enforcement, then the rules of topology tell us that (a) any point *must* be in a polygon (even if this is the 'outside' polygon), and (b) that it can only be in one polygon. (Strictly speaking, a point can also lie on the border between two polygons, but this is a detail which can be safely ignored for the purposes of explanation – for the design of a real algorithm, it would have to be dealt with in a rigorous manner.)

Given these facts, if we project a ray upwards from a point, then the first line which it crosses must form the border of the polygon which contains the point. This means there are now two stages to our point in polygon test:

1. Find the first link that the ray intersects.
2. This link will lie between two polygons, so we will need to discover which of these contains our point.

As with the original version of the test, the key is the fact that the ray is vertical – this means that if we find all the links which the ray crosses; the first one is simply the lowest. The simplest approach to finding which links intersect the ray is to test them all, but we can speed up this process if we calculate the MBR for each link, and use this to filter out the candidate links first.

Once we have found the lowest link, we know that the polygon containing the point is the one below this link – but how do we find out which polygon is below and which above? Once more topology comes to the rescue as the example in Figure 5.12 illustrates.

In this example, the ray intersects the same link twice; but, by looking at the Y values of these intersections, it is easy to decide which is lower. We know the overall direction of the link because we have stored the from and to nodes – the direction is shown with an arrow on the diagram. This means we can tell whether the part of the link where the ray intersects runs from east to west or west to east – in this case, it is EW as the X value of the start of the segment is greater than the X value of the end. With an EW segment, the polygon on the left (C in this case) will be below the segment, and thus will be the polygon we want.

Note that in this case we do not have to count intersections at all. Our topological rules of planar enforcement tell us that we have a set of polygons covering the area (including the 'outside' polygon) and so the point must lie in one of them. The problem is simply to determine which one.

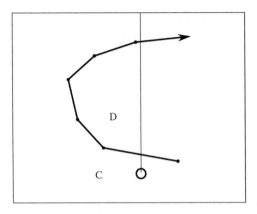

FIGURE 5.12
Testing which polygon is below a link.

Hopefully, this discussion of vector data structures and algorithms has given you a good understanding of some of the issues which arise in writing a piece of GIS software. We have seen that algorithms for apparently simple tasks can be very complicated. We have also seen that the data structure which is used also has a key role to play in algorithm design, and this is an issue which will arise repeatedly in the rest of the book. Throughout the chapters on vector algorithms, there has been a concern with the speed of execution of algorithms, and we have seen some of the strategies which have been used to make algorithms faster. So far, we have not made any formal assessment of how much improvement we can expect from these strategies, and so we will turn to this in the next chapter.

Further Reading

The calculations for polygon areas are covered in Burrough and McDonnell (1998). A brief account of point in polygon is contained in the NCGIA Core Curriculum Unit 33, while Unit 34 covers the polygon overlay operation (Goodchild and Kemp 1990) Both De Berg et al. (1997) and Worboys (1995) give details of other algorithms for the point in polygon problem, and describe algorithms for polygon overlay. Huang and Shih (1997) compare a number of different points in polygon algorithms concluding that the characteristics of the polygons themselves have a strong effect on how quickly the algorithms run.

White (1978) gives some background to the WHIRLPOOL algorithm, which was the first efficient implementation of polygon overlay in a vector GIS. Teng (1986) discusses the importance of both topology and the preprocessing of data in handling the polygon overlay operation.

6

The Efficiency of Algorithms

6.1 How Is Algorithm Efficiency Measured?

We have seen that an important part of designing GIS algorithms is to make them run as quickly as possible. In the case of the simple point in a polygon test, for example, testing the point against the minimum bounding rectangle of the polygon could save a lot of unnecessary line intersection tests. A lot of things can affect how quickly an algorithm will run such as the speed of the computer and how many other tasks are running on it at the same time. However, many of these will be difficult to measure; so, in assessing how efficient an algorithm is, computer scientists concentrate on what is called algorithm complexity – how the performance of the algorithm is affected by the amount of data to be processed. For example, if we have twice as much data to process how much longer will the algorithm take to run? Twice as long? Four times as long?

To see how algorithm complexity is assessed, it is easiest to consider an example of an everyday task – looking up a name in the phone book. The simplest method would be to start at the beginning of the book and work our way through until we found the name we want. If we are lucky, we may find the name quickly – if we are unlucky we may go through the whole book. Let us call this the 'brute force' algorithm. If there are 64 entries in the book, then on average we will search half of them before finding the correct entry – this means making 32 comparisons between the entry and the name we are looking for. If the size of the book doubles, the average number of searches also doubles. The complexity of this algorithm is therefore directly related to the size of the problem. There is a standard notation used by computer scientists to express this so-called 'big-O notation'. In this case, we would say that the brute force algorithm has $O(n)$ computational complexity – the complexity is directly related to the size of the problem, denoted conventionally by the letter n. Capital O stands for order and we could equally say that the algorithm has order n computational complexity. Note that $O(n)$ does not mean it will always take n operations to solve a problem – even with the brute force algorithm, we may get lucky and find the answer straightaway. It is used to indicate the complexity of the problem under the worst case.

The brute force algorithm is a poor algorithm because it does not take advantage of the fact that the names in the phone book are in alphabetical order. Assuming the name we were seeking was 'Wise', most people would start by turning to the back of the book since W is at the end of the alphabet. Once they found the W section, they would look in the first third, since 'I' comes about a third of the way through the alphabet and so on – in effect, rather than searching the whole book, the best approach is to successively narrow down the search to the right part of the book.

To program a computer to perform the same sort of task, we would use what is called a binary search. Rather than start with the first entry in the list, we begin with the middle one – with a list of 64, this would be number 31 or 32 (with an even number of entries there is no middle one, but we would devise a simple rule to pick one of the two entries which straddle the middle of the list). We compare the name with 'Wise'. If they match, the job is done. If 'Wise' comes later in the alphabet, we know we need to look in the second half of the list and vice versa. We therefore repeat this process with the appropriate half of the list. At each stage of the search, we split the list in half, and identify which half the item we are looking for is in. We start with 64 entries, which are split into two groups of 32. One of these is split into 16, and then 8, 4, 2 and 1 – by the time we have only one entry, we will either have found the name we are looking for or know that it is not in the list. This means that the maximum number of comparisons we will have to make is just six. If the size of the list doubles, the maximum number of comparisons only rises to seven – one more subdivision of the list into two halves.

The number of comparisons for a binary search is related to the number of times we have to divide the list into two halves before we are left with a single entry. If the number of items is n, this is the same as the number of times we have to multiply 2 by itself in order to get n. In the first example, n was 64, which is $2 \times 2 \times 2 \times 2 \times 2 \times 2$ or 2^6 (two to the power of six). We had to divide the list into 2 six times at most to find our entry. If n is doubled to 128, this means the number of searches goes to 7 since 128 is 2^7. The complexity of this algorithm is therefore related to the number of times we have to multiply 2 by itself to get n. Fortunately, there is a simple mathematical term for this – the logarithm to the base 2 of n. In mathematical notation,

if $n = 2^m$,

$m = \log_2(n)$

So, the complexity of the binary search algorithm is $O(\log_2(n))$. Logarithms can be calculated using any base – logarithms to base 10 used to be widely used to perform complex calculations in the days before the pocket calculator. However, in terms of assessing the efficiency of different algorithms, the base used for the logarithms is not important (for reasons outlined in Section 6.3 of this chapter) so the efficiency of the binary search is reported as $O(\log n)$.

The binary search algorithm will only work if the list is in order. This is in alphabetical order for our telephone book example, but if we were searching other items, we might need the items in numerical order or date order. So, to make it a general-purpose algorithm, we should really include the first step of checking whether the list is ordered and sorting it if not. Checking is an O(n) operation. We start at the first item and see whether it is smaller or larger than the second – let us assume it to be smaller. Then, we look at the second item and check whether this is smaller than the third and so on down the list. If we find any item out of sequence, the list is unordered, and to discover this we look at every item except the last (which does not have anything below it in the list to be compared with).

To see how we might sort the list, first of all, imagine a related problem. Assume we have an ordered list, and we want to add one more item to it. To find out where it should go, we perform a binary search. If the item is already in the list, this will return its position. If it is not in the list, we will find the position where it ought to go. To sort a complete list, we use the same approach. We begin with our unordered collection of items, and we set up an empty list. We then add the items to the list one at a time. In the first case, the first item is the only thing in the list, so the operation is trivial. The second item can only go before or after the first, and this is also simple. As the list grows, however, the binary search will be the fastest way of finding the correct place in the sequence for the next item. We know a binary search runs in O($\log n$) time, and since we will have to run one for each item to be added, the total complexity of this algorithm is O($n \log n$).

To find the efficiency of the total algorithm then, we simply add together the three terms:

O(n)   for checking whether the list is sorted

O($n \log n$)   for sorting it

O($\log n$)   for performing a search

Hence, O($n + n \log n + \log n$)

In the function within the brackets, $n \log n$ is the dominant term, and so the complexity of the overall algorithm is reported as O($n \log n$). It may seem as if we are ignoring important details in only reporting the dominant element in the function. However, the difference between an algorithm which performs $n \log n$ operations and one which performs $n \log n + \log n$ operations is much less important than the difference in performance between an O($n \log n$) algorithm and one which is O(n^2) as shown in Figure 6.1. This shows how the value of some of the common functions changes as n changes, and also the terms used for some of the functions.

Big-O notation is not only used to assess the processing efficiency of algorithms but is also applied to any aspect of algorithms. When we look at raster GIS later in the book, we will see that an important aspect of different ways

n	Linear $O(n)$	Quadratic $O(n^2)$	Logarithmic $O(\log n)$	$O(n \log n)$
1	1	1	0	0
10	10	100	1	10
100	100	10000	2	200
1000	1000	1000000	3	3000
10000	10000	100000000	4	40000

FIGURE 6.1
Value of some of the functions commonly used in reporting algorithm complexity.

of storing and processing raster data is the trade-off between the amount of memory needed and the speed of answering queries, and both can be assessed using big-O notation.

6.2 Efficiency of the Line Intersection Algorithm

Let us look at the application of big-O notation by assessing the efficiency of the line intersection algorithm.

Figure 6.2 shows two lines, each represented by a series of line segments – 9 in the case of line 1 and 13 in the case of line 2. In Chapter 4, we saw what was necessary to determine whether two line segments intersected. Many GIS operations require the system to compare two sets of lines and determine which of them intersect and where – note that since lines can curve, two lines may actually intersect more than once along their length. The brute force approach to this problem is to test every line segment against every other line segment – if there are n segments, we will perform n^2 intersection tests and this algorithm is of $O(n^2)$ computational complexity.

If we use a minimum bounding rectangle to identify candidates for intersection, then the number of intersection tests will be greatly reduced. It is

FIGURE 6.2
Line intersection example.

FIGURE 6.3
Line intersection using minimum bounding rectangle (from left to right) worst case, best case and average case.

difficult to say exactly how many intersection tests we will have to perform because it depends on the configuration of lines as shown in Figure 6.3. At one extreme, we might have a set of lines in which all the MBRs overlap, and we still have to perform n^2 intersection tests. At the other extreme, we may a have a set of lines in which none of the MBRs intersect, and we have to perform no intersection tests at all. These worst- and best-case scenarios are extremely unlikely in practice of course. A more typical scenario is shown on the right of Figure 6.3, in which each line is near to a small number of other lines.

In the typical scenario, there are two ways in which the number of potential intersection tests could increase. We could be dealing with a larger study area, with lines at approximately the same density – in this case, the number of tests would simply increase as n increased. Alternatively, we could be dealing with a similar-sized area, but containing more lines. Each line would therefore have more neighbours and if the number of extra neighbours was roughly related to the number of extra lines, we might have $2n$ tests to perform.

However, our overall line intersection algorithm is still O(n^2) because although the MBR test reduces the number of full line intersection tests we peform, we still have to compare every MBR with every other MBR, and this is n^2 comparisons. This is another reason why the Bentley–Ottman algorithm, which was described in Chapter 4, is a more efficient algorithm than one based on comparing MBRs. There are two ideas which make the difference. The first is the use of a plane sweep. This is the imaginary line that passes across space identifying which lines should be considered as candidates for intersection at any given time. In Chapter 4, the line was shown sweeping from the left to the right, which means that lines with very different Y values will be potential candidates for testing at any given time. However, this is where the second idea comes into play because the lines under consideration are maintained in the list, which is sorted by the Y coordinate and only lines which are neighbours on this list are compared. If there are n lines on the list, comparing each one with every other one would be O(n^2), but comparing each with 1 or 2 neighbours is O(n). The full algorithm, which is described in de Berg et al. (1997), makes use of two sorted lists. The first is the list of events at which the plane sweep intersects with the end points of the lines and the second is the sorted list of lines under consideration at each stage. Sorted lists can be searched and updated using binary search methods, as described at the start of this chapter, and tend to produce algorithms with O(log n) efficiency. In fact, in this case, testing for all potential intersection can be done in O(n log n) time. However, the total efficiency also depends on how many intersections there are, since each one will require some calculation. In the extreme case of there being no intersections at all, the overall efficiency is O(n log n). More generally, if I intersections are found, the efficiency is O(n log n + I log n). Algorithm efficiency is used calculated on the basis of the size of the input, conventionally represented as n. Algorithms, where the efficiency also depends on the results, are called output-sensitive as their efficiency depends on the output as well as the input.

6.3 More on Algorithm Efficiency

We have seen how big-O notation can be used to identify weaknesses in algorithms and distinguish between efficient and inefficient ones. We will complete this chapter by providing a little more detail on what this notation means, why we are able to ignore some terms in reporting efficiency, and introducing some related measures of efficiency.

In Section 6.2, we saw that when we say that an algorithm has O(n^2) efficiency, this means that the relationship between the number of operations the algorithm performs and the problem size is approximately quadratic. To understand what this means, let us take the example of the binary search algorithm, which was analysed in Section 6.1. We saw that the number of operations needed to find an item in a potentially unsorted list was n + n log n + log n. However, we reported that the efficiency of the algorithm as O(n log n). If we tabulate the values of these two functions for a few values of n, we will see that n log n is always smaller than n + n log n + log n (Figure 6.4). So, it seems as if we are exaggerating the efficiency of our algorithm.

However, we are really most interested in large values of n. Consider our searching problem again. If we have to find a name in a list of five names, it does not much matter how we go about it. The problem is so small that even an inefficient method will be quick enough. If we have 5000 names, the situation is different, however. It is worth the extra effort of sorting the names into order because this will greatly speed up the search process. Sorting five names into order would take almost as long as simply going through the list from 1 to 5 looking for the name we want, and actually make the process slower rather than quicker.

If we look at the numbers in Figure 6.4, we can see that for small values of n, the full cost of our search algorithm (n + n log n + log n) is proportionally

n	$\log n$	$n \log n$	$n + n \log n + \log n$	$2(n \log n)$	$0.5(n \log n)$
2	1	2	5	4	1
4	2	8	14	16	4
8	3	24	35	48	12
16	4	64	84	128	32
1024	10	10,240	11,274	20,480	5120

FIGURE 6.4
Example to illustrate why $n + n \log n + \log n$ is O($n \log n$). Logarithms are to base 2 in all cases.

much greater than $n \log n$ – 2.5 times greater when n is 2. However, as n becomes large, the difference becomes proportionally much less important, so that when n is 1024, the full number of operations is only 10% greater. Big-O notation summarises this situation by saying that if an operation is O($n \log n$), the number of operations it requires is always less than $n \log n$ times a constant factor, for all values of n above a certain value. Column 5 of Figure 6.4 shows the value of $2(n \log n)$. It can be seen that for all values of n above 2, this is greater than $(n + n \log n + \log n)$. This is also very clear from the graph of these numbers in Figure 6.5. In other words, in assessing the number of operations for a search, the function is dominated by the $n \log n$ term, except for small values of n.

This means that in assessing complexity, factors which are either constant or small relative to the main factor are left out. This explains why

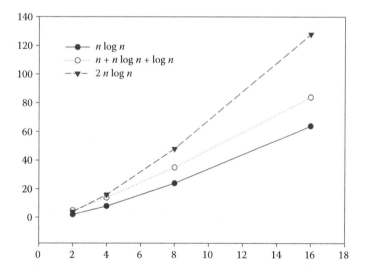

FIGURE 6.5
Graph of three functions of n.

in Section 5.1, the base of the logarithm was unimportant in reporting the complexity of a binary search. The reason is that a logarithm to one base can always be converted to a logarithm in another base by multiplying it by a constant factor (the logarithm of the first base to the second base in fact). For example, if m is $\log_2(n)$, then $\log_{10}(n) = \log_{10}(2) \cdot m$, where $\log_{10}(2)$ is simply 0.3010.

It is also of interest to know whether there are situations in which the algorithm might need far fewer than $n \log n$ operations. In the case of the search operation, this will occur when the list of items is found to be sorted. Checking for this takes n operations, and doing the search takes log n, so the efficiency in this case would be $\Omega(\log n)$. The Greek symbol omega is used to indicate that this is a lower bound for this algorithm – in other words, it can never takes less than $c \log n$ operations, where c is a constant value. What about the search with unsorted data – can this ever do better than $O(n \log n)$? The answer is no because as it stands it uses a sorting algorithm which will always take $n \log n$ operations. This is shown in Figure 6.4, which shows that if we multiply $n \log n$ by a different constant value (0.5 in this case), this will always be less than the number of full operations needed for the algorithm. This means that $n \log n$ describes both the upper and lower bounds for this algorithm, and this is expressed as $\Theta(n)$ using the Greek letter theta.

Since this book is intended simply as an introduction to the ideas of algorithm design, we will not concern ourselves with the distinction between big-O, omega and theta notation, and will simply use big-O to describe the efficiency of different algorithms. This is quite common in the GIS literature, where the concern is normally with how well an algorithm will perform in the worst case, or in the average case, and so the lower complexity bounds are of less interest than the upper bounds.

Further Reading

Worboys (1995) contains a brief explanation of the measurement of algorithm complexity, and an introduction to big-O notation. Any introductory textbook on algorithms, such as Cormen et al. (1990), will contain a more extended discussion on the different forms of notation used for analysing efficiency. Sorting is a classic computer science problem, and was the subject of one of the key publications in the computer science literature – *Sorting and Searching* by Donald Knuth. First published in the 1970s and recently revised (Knuth 1998), this is volume three in *The Art of Computer Programming*, a series which has been voted one of the 12 best scientific monographs of the century by *Scientific American* (ranking alongside the works of Einstein, von Neumann, Feynmann and Mandelbrot). Knuth is also the person who standardised the use of big-O notation. This web page has nice animations of a selection of algorithms operating on data in different initial conditions – http://www.sorting-algorithms.com/. This Wikipedia page is also a good starting point for further exploration of this topic – http://en.wikipedia.org/wiki/Sorting_algorithm.

7

Raster Data Structures

In Chapter 2, we saw that using a raster GIS we could store a set of spatial data in the form of a grid of pixels. Each pixel will hold a value which relates to some feature of interest at that point in space. In fact, there are two main types of raster layer which are used in GIS:

- *Raster imagery*. These days we are surrounded by imagery in digital format – modern televisions and computers all have screens in which the picture is composed of pixels and the photographs we all take on our cameras and phones are in a raster format. We are also used to seeing maps and satellite images used on web pages and even on phones. In all these cases, what is stored is a raster grid of colours, coded in some numerical fashion.

- *Raster data*. We can also store data which we wish to analyse in a raster format. Probably the commonest example is the digital terrain model in which each pixel contains a value for the elevation of the land surface. In general, the values in raster data layers are normally one of three possible types:

  1. Binary – a value which indicates the presence or absence of a feature of interest. For example, in a layer representing roads, we might use 1 for pixels that contained part of a road, and 0 for pixels that did not.

  2. Enumeration – a value from some classification. For example, a layer representing soils might contain codes representing the different soil types – 1 for Podsols, 2 for Brown Earths and so on. Since the values are not directly related to the soil type, there would have to be a key of some sort indicating the meaning of each value.

  3. Numerical – an integer or floating point number recording the value of a geographical phenomenon. In the soil example, we might have measurements of soil moisture content. A common example of this kind of raster layer is when the values represent the height of the land surface, in which case the layer is often referred to as a digital terrain model – these will be described in more detail in Chapter 9.

There is also one important category of data which straddles this simple subdivision, namely, 'remote sensing imagery'. In a remote sensing image, each pixel will usually have a value which is the radiation at a number of different wavelengths and so a pixel can have a large number of values. These are sometimes processed as data to classify the image, for example, and sometimes combinations of wavelengths are treated as colours to produce pictures of the Earth's surface.

The raster data model has the great virtue of simplicity, but it can produce very large files. This is especially true of raster imagery in which each colour is often described in terms of the amounts of the three primary colours, red, green and blue. In what is called 'true colour' representation, the three RGB values can range from 0 to 255 giving over 16,000,000 possible colours. However, this means that each pixel needs three 8-bit bytes to store its colour so a picture displayed on a modest 1024 × 768 pixel PC screen will require over 2 MB of data. Raster data layers tend to have only one value per pixel instead of three, but if these values are real numbers (as can be the case with a DTM), then each pixel actually requires 4 bytes of storage. The file size is also affected by the spatial size of the pixel. This affects the precision with which the data are represented since you cannot represent anything which is smaller than a pixel. This means that pixel sizes need to be small, but the result is very large raster grids. For example, a single digital elevation model (DEM) tile from the British Ordnance Survey's PANORAMA datasets represents an area of 20 km by 20 km with pixels of 50 m – this is 160,400 pixels (401 × 401) and only represents a small percentage of the land surface of Great Britain. What is more, for many applications, even smaller pixel sizes are desirable but halving the pixel size would increase the number of pixels by a factor of 4.

With raster data therefore a key issue is whether it is possible to reduce the amount of data needed to store each layer. We will start by considering this issue in relation to raster data layers and then go on to look at the storage of raster imagery.

7.1 Raster Data in Databases

As the size of raster datasets is a key issue, then the case for trying to handle them using database software is possibly even stronger than it is for vector data since databases are designed to be good at handling large volumes of data efficiently. The difficulties of storing both vector and raster data in normalised tables in a relational database were described in Chapter 2. In Chapter 3, we saw that the potential advantages of storing vector data inside a database have led to numerous developments, including a new version of the SQL standard which can deal with some aspects of storing and querying vector data. So, has the equivalent also happened for raster? Yes and no.

To understand why the situation is different with raster data, we need to understand that as a result of their origins in business computing, databases are designed to hold information about real, practical objects. The first stage in designing a relational database is to build a conceptual model of the information which is going to be stored and processed and this is done by identifying the following:

- Entities to be represented
- Relationships between them

As an example, recall the theatre booking example which was used in Chapter 2. Three of the entities in that example were

- Plays
- Directors
- Performances

and the relationships between them could be defined as follows:

- Each play has one director.
- Each director can direct several plays.
- Each play can have several performances.
- A performance will only be of one play.
- There is no direct relationship between directors and performances.

On the basis of this information, it is then possible to decide what tables are needed and what information they should contain in order that the relationships can be used to link the tables. For instance, if PLAYS and DIRECTORS are held in separate tables, they will need to have something in common so that it possible to perform a join and discover the director for any given play.

This model of entities and relationships works well for vector data where we have discrete points, lines and areas. But, what exactly is the entity in the case of a raster layer? There are really only two things that are possible candidates:

- The whole layer
- The pixels

Treating the whole layer as an entity certainly makes sense, and in fact this has been the basis for the way in which raster data have been handled in GIS databases in the main. If we think of a typical raster dataset such as the digital terrain model in Figure 7.1, we can see that taken as a single object it has a number of attributes which could be well handled in a database.

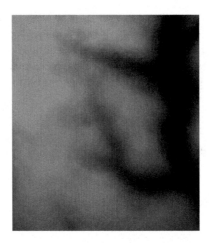

FIGURE 7.1
Example raster layer. (© Crown Copyright Ordnance Survey, all rights reserved.)

There is the geographical extent of the image, the source of the data, the date it was created and who created it, all of which could be handled very easily in a standard relational database table (Figure 7.2).

In fact, if we use a version of an RDBMS which uses the new spatial standard, we can regard the geographical extent of the layer as a polygon, which means we could do spatial searches using this layer with other layers in the database. However, what we cannot do is do anything with the actual pixel values because these cannot be stored as attributes in any useful way. In fact, systems which allow the handling of raster data in databases simply store the pixel values, but if any analysis of them is required, they pass the data onto another piece of software to undertake this.

To analyse raster data, we need to be able to access the values for individual pixels. To take a simple example, to produce a hillshade image from this DTM, we need to know the value of the eight neighbours of each pixel and to be able to do calculations using the values. In theory, it might be possible to produce an extended version of SQL, which allowed this type of query on pixels. However, one of the advantages of fourth-generation languages such as SQL is that they provide a simple and intuitive way for users to handle their data. Doing raster analysis at the level of individual pixels is a little like programming a computer using assembly language – it is too detailed

NAME	Georeference	Xmin	Xmax	Ymin	Ymax	Zmin	Zmax
Slapton	OSGB	273995	287005	38995	55005	10	210

FIGURE 7.2
Some of the attributes for the DEM in Figure 7.1 in a relational table.

and low level. Some work is being done to try and provide an appropriate model for raster data, which would allow it to be analysed in a database. For example, Xie et al. (2012) have developed a version of Tomlin's map algebra which can run inside an ORACLE database. Map algebra certainly provides a good way for users to think about raster analysis and has been the basis of raster analysis tools in many GIS systems. However, interestingly, the approach taken by Xie et al. (2012) is not to try and produce a modified form of SQL, but to extend the third-generation programming language, which is also part of the ORACLE product.

So, currently, databases are routinely used to store and manage raster layers, but not to analyse them. For that, we still need to turn to purpose-written software and to data structures which are specifically designed for the task.

7.2 Raster Data Structures: The Array

The simplest method of storing a raster layer in the memory of the computer is using a data structure called an array. All programming languages have a structure called an array, which can be used to store and access lists of items. In Chapter 6, we considered alternative methods of searching through a list of entries in a telephone book to find one that matched with a particular name. The full list of names to be searched could be stored and accessed using an array as follows:

```
1. Array NAMES[1..64]
2. Read names from file into names array
3. i = 1
4. Found = FALSE
5. repeat until Found == TRUE or i > 64
6.     if NAMES[i] == 'Wise' then Found = TRUE
7.     i = i + 1
```

This is the brute force algorithm for searching the list. The first line sets up an array with space for 64 names and the actual names are read from a file into this array. At this point, the first few elements of the array might contain the following:

NAMES[1]	Smith
NAMES[2]	Jones
NAMES[3]	Wise

Lines 5–7 go through this array, one item at a time, comparing the value with the name we are looking for – Wise – until either this is found or the entire list has been searched. Each element in the array is identified by a

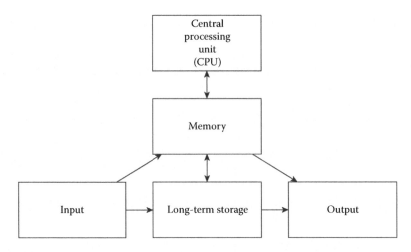

FIGURE 7.3
Schematic diagram showing the main elements of a computer.

number and this number is used to retrieve the correct element from memory. The array is available in programming languages because there are so many cases where it is necessary to deal with collections of related pieces of information. It is also possible to have arrays which have both rows and columns, and these are what can be used to store raster data.

The array is also an extremely efficient data storage mechanism; however, to understand why, it is necessary to look again at the way a computer operates. In Chapter 1, a simple model of the structure of a computer was presented, which is repeated in Figure 7.3.

Recall from Chapters 1 and 2 that computers are programmed to break down a problem into a series of small steps which use only 1 or 2 data values at a time. These values need to be fetched from memory and any results stored in memory. In addition, data will also have to be transferred between long-term storage, normally on disk and memory. The way that memory is organised is that information is stored in fixed length units of bytes or words and that each of these has an address which is simply a number beginning at 0 and going up to the number of bytes or words in the memory. So, our list of names might be as shown in Figure 7.4 when stored in memory.

The circuitry in the computer is designed so that the information can be retrieved from any memory locations equally quickly by passing the address

Address	12905	12906	12907
Contents	Smith	Jones	Wise

FIGURE 7.4
Storage of an array in computer memory.

to the CPU. It is rather as if the postal service worked by having a direct connection between every individual house and the post office.

What a computer program has to do therefore is work out the memory addresses which contain the data which it needs. In the case of the array, this is extremely simple. When an array is set up in a program, the program takes a note of the address of the first element – in this case, the name 'Smith', which is stored in box 12905. The addresses of any of the other elements can then be worked out from the index number, which is normally given in brackets after the name of the array. So, when the program refers to NAMES [2], what the computer actually does is as follows:

```
1. The index value of this element is 2.
2. The first element in this array has an index value of 1.
3. This element is therefore (2-1) = 1 box on from the start.
4. The address of the first element is 12905.
5. The address of this element is therefore 12906.
```

This may seem long winded, but the computer only has to do two calculations – find how far along the array this element is (step 3) and use this to work out the actual address (step 5) – one subtraction and one addition. The calculation in step 3 produces what is sometimes called the offset – how far the element is from the start of the array. In many programming languages, the first array element is labelled 0 so that it is not necessary to perform the calculation in step 3 – the offset is simply the number of the element and this can be added directly to the start address.

To see how this relates to GIS, let us consider the simple raster layer shown in Figure 7.5. Note that for clarity the pixel values are shown as letters, which will help to distinguish them from the numerical memory addresses in the explanations that follow. In practice, most GIS systems only allow the storage of numerical data in pixels.

Instead of a list of names, we now have a set of rows and columns. When we want to identify a particular element in the array, we will need to give both a row and column number – for instance, IMAGE [3,3] to refer to the element in the top right-hand corner. So, does this mean we need a special form

3	A	A	A	A
2	A	B	B	B
1	A	A	B	B
0	A	A	A	B
	0	1	2	3

FIGURE 7.5
Example of simple raster array.

Address	0	1	2	3	4	5	6	7	8	9	10	11	12	13	14	15
Value	A	A	A	B	A	A	B	B	A	B	B	B	A	A	A	A

FIGURE 7.6
Storage of array in computer memory.

of memory which can handle two-dimensional data and two sets of memory addresses – one for rows and one for columns? The answer is no in both cases – we still store our array in a sequence of memory locations in exactly the same way as for our list of names, as shown in Figure 7.6.

To do this, we have to decide what sequence we will use to read the values from the rows and columns into memory. In this case, we have started in the bottom left-hand corner, and proceeded from left to right along each row in turn until we reach the top. This will make the explanation of some of the other ways of storing raster data a little simpler, but in practice many GIS and image processing systems start at the top left and work their way down. There is no single agreed convention, however, and most GIS and image processing systems contain commands to 'flip' raster images which have been read in from systems which use a different convention.

So when a program refers to a particular pixel, such as IMAGE [2,3], how does the computer know which memory location to go to? The size of the array will have been stated at the start of the program. In our pseudocode notation, for example, a four-by-four array would be declared as follows:

```
Array IMAGE[0..3,0..3]
```

Notice that the rows and columns are both numbered from 0 as explained earlier. As before, the program knows the address of the first item in the array – the pixel in the lower left-hand corner. It also knows how many pixels are in each column. So, if we count along two complete rows, and then three pixels along from the start of this row, we will be able to work out the address of the array element we want.

```
address = (nrow*rowsize) + ncolumn
address = (2*4) + 3 = 11
```

You may like to check for yourself that IMAGE [2,3] is the 11th array element starting from the lower left-hand corner.

Note that this calculation is not explicitly performed by the person writing a program in a language such as FORTRAN or C, who declares and uses arrays simply by putting the row and column positions in brackets after the name of the array. When the program is translated into an actual executable program by the compiler, one of the things which is done is to translate these statements into the sequence of operations which will calculate the address of the item and transfer it from memory to the CPU.

One important feature of this address calculation is that no matter how large the array held in memory, the retrieval of an item from it will take exactly the same amount of time. This is indicated by saying that the operation takes O(1) time – the speed is the same, no matter how large the problem. Arrays can therefore be very efficient in terms of processing time. However, they are very inefficient in terms of storage, as every single pixel takes one element of storage. To assess the storage efficiency of various methods of handling raster arrays, it is easy to think in terms of the number of rows or columns rather than the total number of pixels. For any given geographical area, this is determined by the resolution of the raster layer – halve the resolution and the number of rows and columns both double. However, the total number of pixels goes up by a factor of 4. This means that the array, which stores every pixel, has $O(n^2)$ storage efficiency, which is not very efficient at all. To solve this problem, various strategies have been adopted as we shall see in the next few sections.

7.3 Saving Space: Run Length Encoding and Quadtrees

Let us look again at the model of a computer in Figure 7.3. We know that whatever data the CPU needs is passed from long-term storage (usually on a disk) to memory and from there to the CPU, with results being passed back in the opposite direction. In some cases, the source data may come from a different computer entirely across a network, so that the box labelled 'input' may represent anything from numbers typed on a keyboard to a dataset downloaded from a server on the other side of the world. The fastest data transfer occurs between memory and the CPU. Transfer between disk storage and memory is slower than this, partly because of the way data are stored on disk and partly because a disk is a mechanical device. Transfer across the network is slower still because of the number of stages the data have to go through to get from one computer to another and because of traffic on the network.

The main disadvantage of the array as a means of storing raster data is that files stored in this way will be large. Even with modern computers with enormous amounts of disk space and memory, it still makes sense to reduce data sizes for a number of reasons. First, it speeds up data transfer from disk. Second, more images can be held in memory at one time. GIS analysis often involves viewing or using several layers – it is much slower if every time a new one is selected, a file has to be moved out of memory to make way for it. And finally, if we can analyse the data in its compressed format, this will speed up execution times.

The simplest strategy for reducing file sizes is to use the smallest possible amount of storage for each pixel. We saw in Chapter 4 that for the storage of floating point numbers there is a need to store both a mantissa and exponent, and as many digits as possible. For this reason, floating point numbers are held in memory using at least 32 bits and often more. However, the same is not true of integers. An integer is held by converting the number from base 10 to base 2 and then storing the base 2 number. A single byte, with 8 bits, allows for a maximum integer of 255 as shown in Figure 7.7. If one of the bits is used to indicate the difference between positive and negative numbers, then the range is from –128 to 127. Either of these is sufficient to hold the data in many raster layers, which often use small integer numbers – for instance, Boolean layers only use the values 0 and 1 to indicate false and true, respectively. Indeed, these could be held using a single bit, but this is not normally an option which is available. However, the use of single-byte integers is commonly available, and where appropriate will reduce the file size, and hence memory usage by a factor of 4 compared with using 32-bit words.

A second strategy for dealing with large files is to hold only part of the layer in memory at any one time. To assess the efficiency of this approach, we have to consider two issues – how much memory will be needed and how many times will we have to transfer data between memory and disk storage. Assume we have a layer of size n (i.e., with n^2 pixels in total). To process the whole array, we will have to transfer all n^2 pixels between the disk and memory, whether we copy them one at a time, or all at once. However, there is an extra overhead of time every time we ask for a transfer, because the system first has to find the location of the file on the disk, then find the particular part of the file we are requesting. Therefore, we need to try and minimise the number of times we go back and get extra data from the disk.

If we hold the whole array in memory, then this uses $O(n^2)$ storage, but only requires one read and write operation between the disk and the memory. At the other extreme, we could read each pixel as we need it and write it back to the disk afterwards. This now uses only 1 unit of storage but $O(n^2)$ read/write operations. The first option is very quick, but uses a lot of memory – the second uses almost no memory, but would be painfully slow. A compromise is to read one row at a time into memory, process it and write it out to disk – this uses $O(n)$ storage, and also $O(n)$ read/write operations. The difference between these approaches can be quite marked. Wise (1995) describes an example where this was a real issue. The work was on the problem of capturing raster data from scanned thematic maps, such as soil or geology maps. As part of this, a program was written which processed a scanned image, replacing pixels which represented things like text labels,

Binary	Decimal
00000000	0
00000001	1
11111111	255

FIGURE 7.7
Examples of storage of integers in bytes.

3	A	A	A	A
2	A	B	B	B
1	A	A	B	B
0	A	A	A	B
	0	1	2	3

FIGURE 7.8
Example of simple raster array.

lines and so on, with the value for the likely soil or geology category at that point. The program was written for what was then the latest version of the IDRISI GIS, which worked under MS-DOS, and could therefore only access 640 kB of memory. Even with nothing else stored in the memory, the largest image size which could be held in memory would only have been just over 800 columns by 800 rows – in contrast, by processing a row at a time, images of up to 640,000 columns could be handled, with no limit on the number of rows.

These strategies may help, but there are also other things we can do to reduce the size of the image which needs to be stored on disk or held in memory. At the start of this chapter, the main types of values stored in raster GIS layers were identified as binary, enumerated and numerical. In the case of the first two, because the features we are representing occupy regions of the map, the raster layers contain a large number of pixels with the same value next to one another. We can exploit this characteristic to save storage space and the simplest way to do this is to use what is called run length encoding. Consider the simple raster layer we used in Section 7.1, which is repeated as Figure 7.8.

When we stored this as a full array, the first three pixels all contained the same value – A. What we have is a sequence or run of pixels, and instead of storing each one, we can store the information about the run – how long it is and what value the pixels have. Applying this to the whole layer produces the result shown in Figure 7.9.

Even with this small example, we have reduced the number of bytes of storage used for the layer. But, will we always save space in this way? The answer unfortunately is no. Imagine a layer in which every pixel was different from its neighbours, such as a digital terrain model. Every pixel would

Address	0	1	2	3	4	5	6	7	8	9	10	11	12	13	14	15
Value	3	A	1	B	2	A	2	B	1	A	3	B	4	A		

FIGURE 7.9
Storage of a run length encoded layer in computer memory.

take 2 bytes of storage instead of 1 – 1 to record a run length of 1, and 1 for the value itself – so the file size would double.

The final raster data structure we will consider is called the quadtree, and it extends the idea of run length encoding to two dimensions. If we look at Figure 7.8, we can see that there is a block of four pixels in the lower left-hand corner, which all have the value A. Instead of storing four small pixels, it would be far more efficient to store one pixel, which was four times the size of the 'normal' pixel. This is the basis of the quadtree method in which the pixel size is allowed to vary across the image, so that uniform areas are stored using a few large pixels, but small pixels are used in areas of variation.

To illustrate how this works, let us apply it to the layer shown in Figure 7.8. At the first stage, the layer is divided into four quadrants, as shown in Figure 7.10a. Each quadrant is numbered from 0 to 3 in the sequence shown, which is known as Morton order after its inventor. The reason for this particular sequence will become clear later. If we examine each quadrant, we can see that quadrant 0 does not need to be subdivided any further – the values in all the pixels are the same, and so our new pixel 0 can be used to store these data. The three other quadrants are not uniform and so must be subdivided again as shown in Figure 7.10b. Notice that each of the new pixels is labelled by adding a second digit, also in Morton order, to the digit from the first level of subdivision.

As this is only a 4 × 4 image, the process stops at this point. However, how do we store this information in memory, especially now that the pixels are no longer the same size? One method is shown in Figure 7.11.

The first four memory locations (with addresses 0–3) are used to store the results of the quadrants from the first subdivision of the image. The first quadrant, labelled 0, was uniform, and so we can store the pixel value – A. The second quadrant, labelled 1, was not uniform, and so we are going to need 4 bytes to store whatever we found when we subdivided this quadrant. The next available location is at address 4, so we store this address in location 1. As this address is pointing to the location of another piece of information, it is known as a pointer. We have to do the same thing for quadrants 2 and

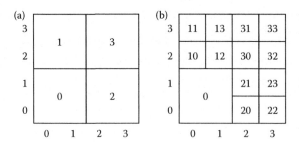

FIGURE 7.10
Quadtree subdivision of layer shown in Figure 7.6. (a) First subdivision and (b) second subdivision.

Address	0	1	2	3	4	5	6	7	8	9	10	11	12	13	14	15
Quadtree address	0	1	2	3	10	11	12	13	20	21	22	23	30	31	32	33
Value	A	4	8	12	A	A	B	A	A	B	B	B	B	A	B	A

FIGURE 7.11
Storage of quadtree in memory.

3, storing pointers to addresses 8 and 12, respectively. The four address locations starting at 4 are used to store the results of subdividing quadrant 1 to produce 10, 11, 12 and 13 – since these were all uniform, we can simply store the pixel values, and in fact this is true for all the remaining pixels.

In this case, we have not saved any space at all compared with the original array method, because there are not enough large uniform areas on the layer. In fact, as with run length encoding, we could end up storing more information. If quadrant 0 had not been uniform, we would have needed an extra 4 bytes of storage to store the individual pixel values making up this quarter of the image. However, in real-world examples, the savings in space can be considerable.

In this example, the image was conveniently the right size to make it possible to divide it into four equal quadrants. When this is not the case (i.e., most of the time) the image is expanded until its sides are both powers of 2, filling the extra pixels with a special value indicating 'no data'. This increases the amount of storage of course, but since the extra pixels are all the same, they can generally be represented using fairly large pixels, and the additional data are more than offset by the savings due to the quadtree itself.

So, why is it called a quadtree? The quad part is obvious, from the subdivision into quadrants. The tree comes from a common way of representing such data structures in computer science as shown in Figure 7.12.

The first level of subdivision is represented as four branches from the original image. Where quadrants are subdivided, four further branches are drawn, giving a tree-like structure. The ends of the lines are all called nodes.

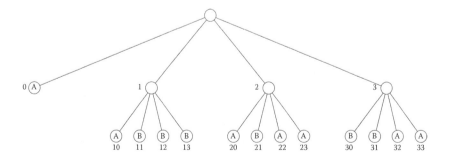

FIGURE 7.12
Graphical representation of quadtree.

Those where the process ends (as in the case of 0 and 10, 11, 12 and 13, etc. in the diagram) are called leaf nodes, while those which represent points of further subdivision are called non-leaf or branch nodes. The first node of the tree is called the root node, even though trees are usually drawn with the origin at the top, as in Figure 7.12.

In Figure 7.11, the Morton addresses of the pixels have been shown, for clarity. In fact, this information is not stored in reality because there is a simple method for calculating it from the row and column number of a pixel as we will see in the next chapter, when we look at how we can use these different data structures to answer a range of GIS queries. We will also come across other tree structures later in the book because they are a very widely used data structure.

7.4 Data Structures for Images

The issue of file size is as important for raster imagery as it is for raster data, but there are some differences in the problems which need to be solved. First, raster imagery will be displayed rather than being used for analysis. This means that however it is stored, it will need to be converted to a full array format because that is what will be needed to show the image on a raster screen. So, why go to all the trouble of trying to make the file smaller if it is going to be reconverted to its full size? There are a number of reasons. One is storage space. Storage capacities increase all the time. When I worked at a university computer centre in the 1980s, a 100 MB disk drive was considered very large. Now, I am storing the word processed files for this book on a 1 TB disk – 10,000 times the capacity. However, the data we wish to store and use also increase in size and so there is a continued need to keep files as small as possible. The other issue is speed. The image will need to be passed from secondary storage into memory to be displayed; as explained earlier, this transfer between storage and memory is relatively slow. The smaller the file, the faster the transfer. This issue is even more important when raster imagery is used across a network. Increasingly, the model of computer architecture shown in Figure 7.3 needs to be modified to show that in many cases a computer will be connected to a network, which means that the box labelled input, which on a standalone PC might be thought of as the keyboard or mouse, could in fact be the input from another computer on the network. Clearly, transmission across a network is going to be much slower than transmission within the computer or to devices which are locally attached. In many cases of course, the network will be the World Wide Web and so the image which is being displayed may havecome from a server halfway round the world. In this context, a lot of thought has to be given as to how to make this process as fast as possible.

Work on data structures for the GIS and for computer images has tended to be carried out in separate academic disciplines. However, interestingly, there is a good overlap in the ideas which have been developed as we shall see as we consider some of the algorithms used in some common image formats.

Both run length encoding and the quadtree are examples of what are called image compression techniques, that is, techniques for making raster data smaller. Both take advantage of the fact that, with spatial data, adjacent values are commonly the same and this idea is used in many image compression techniques. The first we will consider is the GIF or graphics interchange format, which makes use of a general compression algorithm called LZW after the names of its three creators – Lempel, Ziv and Welch. We saw that in run length coding, whenever the same data value was repeated, this was replaced by a count of how long the repeating run was. LZW is based on a similar idea but with two differences. First, the algorithm looks for repeating patterns of values in the input such as ABCABC rather than for repeating single values. Second, the repeat patterns are represented by a code. The easiest way to explain the algorithm is to see it in action. The example which will be used is this very simple data, which simply repeats two data values

ABABABAB

With GIF, it is necessary to know beforehand what values can occur in the data. In this case, we will assume that we know that the only values which can occur are A and B. We therefore construct what is called a data dictionary in which each of these values is allocated a code:

Value	Code
A	1
B	2

The algorithm replaces characters in the original data file with their code value from the dictionary. The compression comes in because as strings of characters are found in the input they are added to the dictionary and replaced by a single code. Each time the string is found later in the input, it is replaced by a code, and because of the way the algorithm works, the strings which can be replaced in this way grow longer and longer as the file is processed.

In simple terms, the algorithm can be described as follows:

```
1. Take one character from the input - call this string S
2. Take one more character from the input - call this T
3. See if S + T is in the data dictionary
4. If it is THEN
       a. Add T to S to form a new S
       b. Go back to step 2
5. ELSE
```

```
        a. Output the code for S
        b. Add S + T to the data dictionary
 6. Go to step 1
```

Let us see how this works for our test sequence – remember that the sequence is ABABABAB.

```
 1. In the first step, S will be A and T will be B. The
    string AB is not in the dictionary, so the algorithm will
        a. Add AB to the dictionary and give it the next free
           code value – 3
        b. Output a value of 1 to represent the string 'A'
 2. In the second step, S will B and T will be A, so the
    algorithm will
        a. Add BA to the dictionary and give it the next free
           code value – 4
        b. Output a value of 2 to represent the string 'B'
 3. In the third step, S + T will be AB, which now is in the
    dictionary. However, adding the next character A gives a
    new string – ABA. So, the algorithm will
        a. Add ABA to the dictionary with code 5
        b. Output 3 to represent the string 'AB'
```

As the algorithm proceeds longer and longer, repeat strings will be found and the amount of data compression will increase. As a result of the way the dictionary is built up, when a GIF file is decoded and turned back into the original string, the dictionary can be reconstructed from the information in the coded file, meaning there is no need to send the dictionary with the file.

Although LZW compression is at the heart of the GIF format, there are lots of other characteristics of this graphics format. One that is of particular interest here is the ability to store the image in what is called interlaced format. This means that instead of storing the first row followed by the second, all the odd rows are stored first, followed by all the even rows. This means that when the image is drawn, it appears in a reduced resolution version first with only half the lines visible. This takes advantage of the fact that human beings are very good indeed at recognizing images and will know very quickly whether the image is the one they want or not. If they do not, then the transmission of the image can be stopped, saving time and saving traffic across the network.

The fact that people are so good at recognizing images also underlies the development of what is probably the commonest image format in use on the World Wide Web, the JPEG, named for the Joint Photographic Experts Group, which created and develops the format. All the data structures which we have covered so far have been concerned with compressing the data, but in a way that preserves all the original information. However, if the image is simply going to be viewed, rather than have its content analysed, then it is not necessary to retain all the detail. Just think of the example of a comic strip in a newspaper. Think of how few lines are used by the artist to draw the characters and yet comics can be used to convey a wide range of situations and emotions. They can do this because people are very good at reconstructing objects in imagery from partial information. If it is possible to simplify an image, while retaining the essentials, then this opens up the possibility of much greater degrees of compression and much smaller files. Compression methods which do not retain all the data are known as 'lossy' and in JPEG the information loss occurs in two main parts of the algorithm. The first comes at the very first stage. As described earlier, colours in imagery are stored using RGB triplets, that is, three numbers with values for the amounts of red, green and blue in the image. However, a lot of the information in an image is still perfectly clear if the image is converted to a greyscale image, with no colours at all as can be seen from the picture of a bridge in Figure 7.13.

The original photograph, which was 2048×1536 pixels was about 9 MB in size. Simply converting from RGB triplets to greyscale, which means that there is now only one value per pixel, reduces this to about 3MB and yet the image is still perfectly clear.

What JPEG does is not to throw away the colour information, but simplify it. To do this, it first converts the image from RGB to what is called YCbCr format in which one number represents brightness, and two the colour. The brightness, or luminance to give it its technical name, is the Y value and this says where the brightness lies on the range between black and white, which is what a greyscale image contains. The other two values, Cb and Cr are both called Chroma values and they encode the colour information. All three values are calculated from the RGB triplet, so it is possible to convert back to RGB from YCbCr. However, since colour is less important for the perception of the image, in JPEG, the Cb and Cr layers are sampled by discarding every

FIGURE 7.13
Greyscale image of bridge over a stream.

second pixel in the horizontal and usually also in the vertical directions. At a stroke, this reduces the size of these two layers by a factor of 2 or 4.

To explain the second element of lossy compression, it is necessary to introduce the idea of spatial data as a signal. Let us start with a more common example of something which can be considered as a signal, namely, sound. Most people know that sound is generated by anything which causes a vibrating movement in the air. Our ears detect the movement and interpret this as sound. The movement can be described in terms of a waveform as shown in Figure 7.14.

The figures represent the same note, A, but played on two different instruments. In both cases, the wave has roughly the shape of a sine curve and has the same frequency of 440 Hz, which means that the curve describes 440 oscillations per second. It is this frequency which determines that what the ear hears is an A. However, as the figures show, the wave is more complex than this and is different on the two instruments. This is because when the string is plucked, as well as vibrating at its main pitch, it also vibrates at related pitches, such as the octaves above and below. It does this at a much lower volume than the main pitch and so the effect is of an A note, overlaid with quieter, related notes known as harmonics. It is the mix of different harmonics which actually gives the different instruments their characteristic sound qualities. It is these harmonics which cause the bumps on the waveforms in Figure 7.14 and which is why the waves are not perfect sine waves. However, it would be possible to remove all the harmonics from the

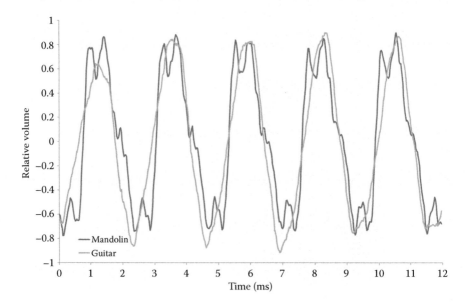

FIGURE 7.14
Examples of sound waves – the note A as played on a guitar and a mandolin.

sound and what would be left would still be an A note – it would just sound rather flat and electronic. The key points are that sound can be described by a waveform, and that this waveform will be composed of a range of frequencies, some more important than others.

So, what is the relevance of all this? The point is that anything which shows variation over time or space can in theory be represented as a wave, and studied in terms of its frequencies. Figure 7.15 shows a very small section of the picture in Figure 7.13 as a greyscale image. If the values in this image are thought of not as colours, but as numbers ranging from 0 to 255, then the data can be treated as a signal. For instance, if we just take the first row of pixels and plot the value against the position on the X axis (Figure 7.15b), we get a graph which can be thought of as a waveform.

The analysis of this kind of data in terms of its frequency characteristics is based on the work of French mathematician Joseph Fourier who discovered that it is possible to approximate any curve as a series of sine waves added together. Each sine wave matches one of the frequencies present in the data, and it is possible to identify the relative importance of each in describing

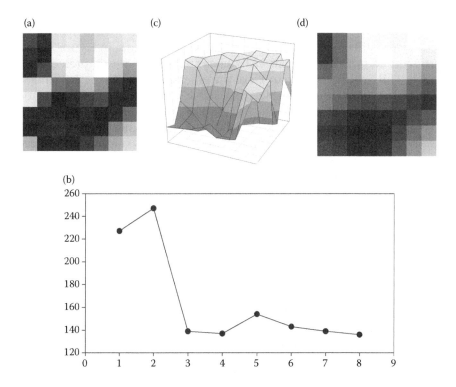

FIGURE 7.15
8×8 pixels from the top left corner of Figure 7.13 represented as a raster layer (a). The first row is shown as a profile (b) and the whole area as a surface (c). (d) The area after one pass of a smoothing filter, which simulates spectral filtering.

the curve. Applying Fourier analysis to our musical examples in Figure 7.14 would show that both have a strong frequency at 440 Hz and then a series of less important frequencies. If we wished to produce an approximation of our waveforms, we simply use the stronger frequencies and ignore the smaller ones.

So, we could analyse the waveform in Figure 7.15, find the main frequency, and simplify the data by approximating it using a sine curve of the same frequency. This is essentially what is done in creating a JPEG except that instead of considering each row of the image, we consider each 8 × 8 pixel block as a surface, as shown in Figure 7.15c, and analyse its frequencies in two dimensions. The effect of simplifying the image by removing the less important frequencies is illustrated in Figure 7.15d. This has not been produced by Fourier analysis, but by a simple averaging; however, the effect of ignoring small variations while leaving the main trends is the same.

So, the second element of the JPEG compression is to identify the frequency variations across each 8 × 8 set of pixels across the image and retain only the most important. How many frequencies are retained is set by the user who can specify how much compression they require. Quite large reductions in file size can be achieved without much loss in picture quality, which is why JPEG has become so popular. Even with very large reductions in file size, the picture may still be recognizable. For instance, the picture in Figure 7.16 has been reduced from the original 9 MB when stored in an uncompressed form to just 60 kB, which is less than 1% of the original size and yet the main elements of the scene are still visible. Also visible is the blocky appearance from splitting the picture into 8 × 8 blocks.

This section of the chapter has introduced some rather new ideas about handling raster imagery, but the final stages of the JPEG compression process

FIGURE 7.16
Picture of the bridge with very high compression applied.

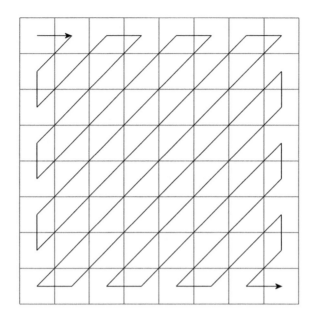

FIGURE 7.17
Zig-zag pattern used to process the coefficients in an 8 × 8 JPEG block.

will reintroduce us to some which we have already encountered in this chapter. The process of analyzing the frequencies in an 8 × 8 block of pixels produces an 8 × 8 array of coefficients, which is what is actually stored. Because of the way the coefficients are calculated, the largest is always in the top lefthand corner of this array, and the coefficients become smaller as you go down and across. In other words, there is a strong spatial pattern in the data and we have already seen that we can take advantage of this fact when trying to compress raster data. One way in which this is done is that the values in the 8 × 8 block are stored in the sequence shown in Figure 7.17, which resembles the Morton sequence in that pixels which are close in space are close in this sequence. This means that a form of run length encoding can be used to group together the very low values which tend to occur together before employing a final method of data compression called Huffman coding to the create the final JPEG file.

Further Reading

Brian Berry's two books (1993, 1995), based on his articles in GIS World, provide a gentle introduction to raster data structures and analysis. The NCGIA Core Curriculum Units 35 and 36 (http://www.geog.ubc.ca/courses/klink/

gis.notes/ncgia/toc.html) cover raster data structures for raster data. The classic texts on quadtrees are the two books by Samet (1990a,b). These describe a range of different quadtree data structures and algorithms for performing standard GIS operations such as overlay using quadtrees. There is now also a website (http://www.cs.umd.edu/~brabec/quadtree/) by Brabec and Samet, which provides applets demonstrating many quadtree algorithms. Try one of the point or rectangle quadtree demos – add some points/rectangles, and then move one to see how the quadtree decomposition of space changes to store the information as efficiently as possible. Rosenfeld (1980) and Gahegan (1989) are both shorter summaries, which provide a good overview, while Waugh (1986) presents some cogent arguments as to why quadtrees may not be a good thing after all! Finally, Dyer (1982) presents an analysis of the space efficiency of quadtrees.

8

Raster Algorithms

The previous chapter described some of the commonest data structures for the storage of raster data and emphasised that at least one of the reasons that they were developed was to produce smaller files for raster GIS layers. Smaller files mean that less disk space is needed to store them and that it will be quicker to transfer them between the disk and the memory of the computer. A small file size should also mean that operations on the layers, such as queries and overlays, should run more quickly, because there are less data to process. However, this will only be true if the query can use the raster layer in its compressed form, whether this be run length encoded or in a quadtree or in any other format. If this is not possible, then the layer would have to be converted back to its original array format (which would take time) and then the query run on this expanded form of the data. So, the data structures which are used for raster data not only have to produce savings in file size, but they must also be capable of being used for a range of GIS queries. To illustrate this, we will consider some simple raster GIS algorithms, and show how these can be implemented using the three raster data structures described in Chapter 7.

8.1 Raster Algorithms: Attribute Query for Run Length Encoded Data

The first is the most basic query of all – reporting the value in a single pixel. To illustrate this, we will use the simple layer in Figure 8.1, and imagine that we wish to find out the attribute value stored in the pixel in row 2 column 3. Remember that rows and columns are numbered from zero.

In a real GIS of course, this query is more likely to take the form of an on-screen query, in which the user clicks a graphics cursor over the required pixel on a map produced from the data. However, it is a simple matter for the software to translate the position of the cursor to a pixel position on the layer itself.

We have already seen how this query would be answered using the array data structure in Chapter 7. Assuming that the layer is stored in an array called IMAGE, the value we want will be held in the array element IMAGE (8.2, 8.3). To retrieve this value from the array as it is stored in memory (Figure 8.2), we

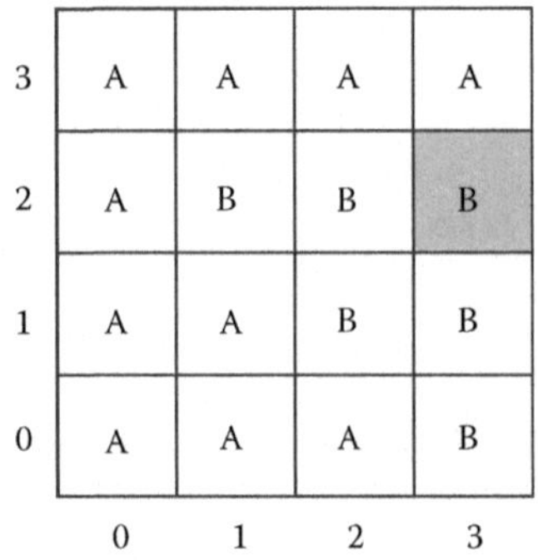

3	A	A	A	A
2	A	B	B	B
1	A	A	B	B
0	A	A	A	B
	0	1	2	3

FIGURE 8.1
Example of simple raster array – pixel in row 2 column 3 is highlighted.

calculate the address from the row and column number by first calculating the offset – how many elements along the array we need to go:

```
offset = (nrow*rowsize) + ncolumn
offset = (2*4) + 3 = 11
```

To find the actual address in memory, we add the offset to the address of the first element, which is 200

```
address = origin + offset = 200 + 11 = 211
```

If we look at Figure 8.2, we can see that the value in address 211 is *B*, which is correct.

Notice that the addresses are different from those given in Chapter 7. This is deliberate because the actual addresses will normally be different every time the software is run. When a program, such as a GIS, is loaded into memory ready to be run, where exactly in memory it goes will depend on what is already running on the computer. If this is the first package to be run, then it will be loaded near the start of the memory (and the addresses will be small numbers). If there are already other programs running, such as a word processor or spreadsheet package, then the GIS will go higher in memory

Address	200	201	202	203	204	205	206	207	208	209	210	211	212	213	214	215
Value	A	A	A	B	A	A	B	B	A	B	B	B	A	A	A	A

FIGURE 8.2
Storage of array in computer memory.

Address	200	201	202	203	204	205	206	207	208	209	210	211	212	213	214	215
Value	3	A	1	B	2	A	2	B	1	A	3	B	4	A		

FIGURE 8.3
Storage of image in run length encoded form in computer memory.

and the addresses will be larger numbers. One of the things which happens when a program is loaded into memory is that the origins of all the arrays are discovered and stored so that all the array referencing in the program will work properly.

If we look at the same raster layer stored in run length encoded form (Figure 8.3), we can see that this same approach is not going to work because there is no longer a simple relationship between the addresses and the original row and column numbers. For instance, the value stored at offset 4 will always refer to the second run found in the original layer, but which row and column this relates to depends entirely on the length of the first run. What we have to do is work out the offset ourselves, and then count along the runs until we find the one that contains the pixel we want.

The program we use to do this will be as follows:

```
1. array RLE[0..15]
2. offset = (nrow * rowsize) + ncolumn
3. pos = 0
4. n = 0
5. repeat until pos >= offset
6.          pos = pos + RLE[n]
7.          n = n + 2
8.       value = RLE[n – 1]
```

The run length encoded storage of the image is declared as a one-dimensional array on line 1. Since we cannot take advantage of the automatic facilities provided by the two-dimensional row and column notation, we may as well treat the image in the way that it is actually stored in memory – as a single row of values. Line 2 calculates the offset, and is the same as the calculation that is done automatically when the image is stored using a two-dimensional array. The key to this algorithm is line 6. The first time through this program, both pos and n will have the value 0 (set on lines 3 and 4). RLE[0] contains the length of the first run (6) so pos will take the value 6. If this is greater than or equal to the offset we are looking for, then this first run would contain the pixel we are looking for. However, it does not, so we go back to the start of the repeat loop on line 5. When line 6 is executed this time, n has been set to a value of 2 (on line 7), so RLE[2] will give the length of the second run – 2. This is added to pos, making 8, which is still less than our offset, so we go round the repeat loop again. Eventually, on the fourth time around, the value of pos will reach 12 – this means, we have found the run we are looking for and the repeat loop ends. By this stage, n has a value of 8, since it is pointing to the length of the next run, so we look at the value in the previous array element to get the answer – B.

It is immediately clear that this algorithm is less efficient than the one for the array. In fact, every time we make a reference to an element in RLE in this example, the computer will have to do the same amount of work as it did to retrieve the correct answer when we stored the data in an array. Answering this query using an array had order O(1) time efficiency – it works equally quickly, no matter how large the raster layer. In the case of the run length encoded data structure, the time taken is related to the number of runs which are stored. Other things being equal, we might expect this to be related to the size of the layer. The layer in Figure 8.1. is 4 × 4 pixels. If we doubled the size to 8 × 8, the number of pixels would go up by a factor of 4. However, the number of runs would not go up quite this much because some runs would simply become longer. The exact relationship would depend on how homogeneous the pattern of values was. A reasonable assumption is that the runs would roughly double in size if n doubled – hence, our algorithm would have O(n) processing efficiency.

8.2 Raster Algorithms: Attribute Query for Quadtrees

We seem to face the same problem in carrying out our attribute query using a quadtree as we did with the run length encoded version of the layer – there appears to be no relationship between the position of the pixels in the original GIS layer, and their position in the data structure. However, oddly enough, this is not the case because when we created the quadtree, we saw that we were able to give each node in the tree a Morton address (shown on Figure 8.4), and there is a relationship between these Morton addresses and the row and column numbers.

Let us take the same query as before – find the value of the pixel in column 3, row 2. To discover the Morton address, we first have to convert the column and row to binary. For those who are not familiar with changing between number bases, this is explained in detail in a separate box – those who are familiar will know that the result is as follows:

$$3_{10} = 11_2$$

$$2_{10} = 10_2$$

where the subscript denotes the base of the numbers. We now have two, 2-bit numbers – 11 and 10. We produce a single number from these by a process

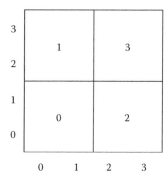 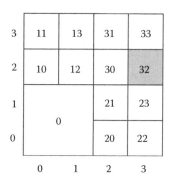

FIGURE 8.4
Quadtree subdivision of layer shown in Figure 8.1, with the pixel in column 3, row 2 highlighted.

called interleaving. We take the first (leftmost) bit of the first number followed by the first bit of the second number, then the second bit of the first number and the second bit of the second number as shown below:

First Bit of 11	First Bit of 10	Second Bit of 11	Second Bit of 10
1	1	1	0

This produces a new number 1110_2. Finally, we convert this from base 2 to base 4, which gives 32_4. Again, this is explained separately in the box for those who are unfamiliar with conversion between different number bases. If you look at Figure 8.4, you can see that this is indeed the Morton address for the pixel in column 3, row 2. So, why does this apparently odd process produce the right answer? The answer lies in the way that Morton addresses were derived in the first place.

One of the problems with spatial data is just that – they are spatial. One of the most important things to be able to do with any kind of data is to put it into a logical sequence because it makes it easier to find particular items. We have already seen how having a list of names in alphabetical order makes it possible to find a particular name using things like a binary search. This algorithm is only possible because the names are in a sequence. But, how can we do the same with spatial data, which are two-dimensional? If we take pixels and order them by their row number, then pixels which are actually very close in space (such as row 0, column 0 and row 1, column 0) will not be close in our ordered list. If we use the column number instead, we will have the same difficulty. This was the problem that faced the developers of the Canadian GIS, and the solution they came up with was to combine the row and column numbers into a single value in such a way that pixels that were close in space were also close when ordered by this value.

Address	200	201	202	203	204	205	206	207	208	209	210	211	212	213	214	215
Morton code	0	1	2	3	10	11	12	13	20	21	22	23	30	31	32	33
Value	A	4	8	12	A	A	B	A	A	B	B	B	B	A	B	A

FIGURE 8.5
Storage of quadtree in memory.

The Morton code, named after its inventor, was the result. Therefore, the reason that quadtree pixels are labelled in this way is because there is then a simple means of deriving their quadtree address from the row and column number. The Morton code is also known as a Peano key, and is just one of a number of ways in which pixels can be ordered. The question of how spatial data can be organised to make it easy to search for particular items is a very important one, and will be considered further in Chapter 12.

Let us return to our original query. The next stage is to use our Morton address of 32 to retrieve the correct value from the quadtree as it is stored in memory (Figure 8.5). As before, in this representation of the data structure, leaf nodes are indicated by a letter, and non-leaf or branch nodes by a number which is a pointer to the location of the first of its child nodes.

Each digit in the Morton address actually refers to the quadrant the pixel was in at each level of subdivision of the original layer, and this is the key to using the address in conjunction with the data structure. In this case, the first digit is 3, which means the pixel was in quadrant 3 when the layer was first subdivided (this can be checked by looking at Figure 8.4). When we stored the quadtree in our data structure, the results for the first subdivision were stored in the first four locations, starting at offset 0. The value for pixel 3 will therefore be stored in an address, which we calculate by

```
Address of start of the array + 3
```

In Figure 8.5, the start of the array is at address 200, so the value we want is at address 203. The value stored in address 203 is number 12. As this is a number, rather than a letter, it means that the results of the second subdivision are stored in the four locations starting at an address which is offset from the start of the array by 12 or 212. The second digit of our Morton address is a 2, so the value we are looking for will be stored in the location which is offset from 212 by 2, which gives us address 214. Looking at Figure 8.5, we can see that the value stored at offset 214 is indeed *B*, the answer to our query. In the way in which the quadtree has been stored here, the pointers within the tree point to an address relative to the root of the tree, whereas the Morton codes give an address which is relative to the current branch of the tree.

What would have happened if we had requested the value for the pixel on row 0, column 1? The Morton address in this case would be 0001_2, or

01_4. Note that we do not leave off the leading zeroes from these addresses. Applying the same algorithm as before, we take the first digit, 0, and look in our data structure. This time we find a letter, A, rather than a number. This means that we have found a leaf, rather than a branch, and in fact this is the answer to our query. There is no need to use the other digits in the Morton address, since all the pixels in quadrant 0 at the first subdivision had the same value.

The pseudocode to implement this algorithm would look something like this

```
1.   Array QUADTREE[0..15]
2.   found = FALSE
3.   repeat until Found == TRUE
4.   Take next digit from address
5.   offset = offset + digit
6.   if QUADTREE[offset] is a LEAF then
7.       value = QUADTREE[offset]
8.       Found = TRUE
9.   else
10.      offset = offset + QUADTREE[offset]
```

As with the run length encoded version, we treat the quadtree as a single array, and find our way to the correct element of the array using an algorithm rather than the in-built system which we had with the two-dimensional array. The efficiency of this algorithm will be related to the number of times we have to go round the repeat loop in the above program. Each time we go round this loop, we are going down one further level of the tree, or one further level of subdivision of the original image. In our simple example, we had a 4×4 image, which was divided into quadrants two times. If we doubled the size of the image to 8×8, this would only add one to the number of subdivisions, which would become 3. You will probably have noticed a pattern here, since

$$4 = 2^2$$

$$8 = 2^3$$

In other words, the maximum number of levels in a quadtree is the number of times we can divide the size of the quadtree side, n, by 2 – in other words, $\log_2(n)$. This means that our algorithm for finding any given pixel has an efficiency of $O(\log n)$.

Before we leave this particular algorithm, it is worth taking time to explore the fourth line in the code sequence above since it affords a useful insight into the way that computers actually perform calculations. Given the Morton address 32, the first time through the repeat loop, we need to select digit 3, and then digit 1 the next time around. But how is this done? Imagine for a moment that we needed to do the same thing in the familiar word of base 10 arithmetic. To find out what the first digit of 32 was, we would divide the number by 10, throwing away the remainder:

$$32/10 = 3.2 = 3 \text{ with remainder } 0.2$$

When we divide by 10, all we do is move the digits along to the right one place – if we multiply by 10, we move them to the left and add a zero at the right. Exactly the same principle applies with binary numbers.

In the memory of the computer, the address 32_4 would be stored as a 32-bit binary number:

$$00000000000000000000000000001110 \qquad (8.1)$$

Note that two binary digits (bits) are needed for each of our base 4 digits – 3 is 11, 2 is 10. If we move all the bits two positions to the right, we will get the following:

$$00000000000000000000000000000011 \qquad (8.2)$$

This is our first digit 3, which is what we need. We have divided by 100_2, but we have done it by simply shifting the bits in our word, an operation which is extremely simple and fast on a computer. In fact, on most computers, this sort of operation is the fastest thing which can be done.

The second digit we need is 2, which looks like this:

$$00000000000000000000000000000010 \qquad (8.3)$$

The bits we need are already in the correct place in the original Morton address (8.1) – all we need to do is get rid of the 11 in front of them. The first step is to take the word containing the value 3 (8.2), and move the bits back two positions to the left giving us the following:

$$00000000000000000000000000001100 \qquad (8.4)$$

This moves the 11 back to where it started, but leaves zeroes to the right of it. If we now subtract this from the original Morton address (8.1), we will be left with 10.

This approach will work no matter how large the Morton address. Assume we have an address which is 1032_4. For simplicity, we will assume this is stored in an 8-bit register called R1 inside the CPU. The algorithm for extracting the leftmost digit is as follows:

Step	Operations	Contents of Registers
1	Make a copy of R1 in R2	R1: 01001110
		R2: 01001110
2	Shift R1 right by 6 bits	R1: 00000001
	R1 now contains the first digit	R2: 01001110
3	Shift R1 left by 6 bits	R1: 01000000
		R2: 01001110
4	Subtract R1 from R2	R1: 01000000
		R2: 00001110
5	Copy R2 to R1	R1: 00001110
	The first digit of the code has been set to 00, and we are ready to extract the second digit	R2: 00001110

At the end of this, R1 contains the original Morton address, but with the first digit (the first two bits) set to zero. To obtain the second digit, we can simply repeat this sequence, but shifting by 4 bits rather than 6. Then, for the third digit, we shift by 2 digits, and the final digit by 0 digits (i.e., we do not need to do anything). The size of the original address was 4 base–4 digits. Given a Morton address of size n, this will be stored in $2n$ bits. To get the first Morton digit, we shift by $2n - 2$, to get the second by $2n - 4$ and so on – so to get the mth digit, we shift by $2n - 2m$ bits.

We have assumed that we already have the Morton address we need but of course we will need to work this out from the original row and column numbers. As these will be stored as binary integers, they will already be in base 2. To interleave the bits, we will also use a series of bit shifting operations to extract each bit from each number in turn, and add it in its place on the new Morton address. As with extracting the digits from the Morton address, it may seem very complicated, but all the operations will work extremely quickly on the computer.

CONVERSION BETWEEN BINARY AND DECIMAL

If we have a number in our usual base 10 numbering system, how do we convert this to a different system, such as binary? First it is important to understand that any number, in any number base, is effectively a sum. Imagine trying to count a flock of 27 sheep on your hands. When you get to 10 you run out of fingers, so what do you do? You might use your toes instead but this will only allow you to get to 20 before you run out of things to count with. Alternatively, you might make a mental note that you have counted 1 lot of 10 and start again. The second time through the fingers you get to 20, make a mental note that this is the second lot of 10 and start again until you get to the last sheep. Your count is therefore 'two lots of 10 plus seven more', which is 27 in total. In other words, the number 27 can be thought of as

$$27 = 2.10 + 7$$

Instead of fingers of course we use symbols to represent numbers – the familiar digits from 0 to 9. The same logic can be applied to counting our flock of sheep. We count up to 9. At this point, we have run out of digits. We could keep inventing new symbols for larger numbers, but this would be very cumbersome. Instead, we record the 10th sheep using the digit 1 – one lot of 10, and start counting from 1 again. If the number of sheep was 271, this method runs out of digits again at 99 – we do not have another digit to record that we have reached the 10th lot of tens. So, we use 1 to record that we have reached 10 lots of 10 for this first time. Therefore, the number 271 can be represented as

$$271 = 2.10.10 + 7.10 + 1$$

Ten times 10 can also be represented as 10^2 (ten squared). Ten itself is 10^1 (ten to the power 1) since any number to the power 1 is simply itself. One is 10^0, since any number to the power nought is 1. So, our number becomes a sum of successive powers of 10:

$$271 = (2.10^2) + (7.10^1) + (1.10^0)$$

In our decimal numbering system, powers of 10 are used because there are 10 digits (0–9) and because the system originated from counting on our 10 fingers (which are also called digits of course – hence the name for the mathematical digits). However, we do not have to use 10 as our number base. When designing computers, it was found that it was far easier to design machines which could use just two digits – 0 and 1 – for storing and processing information. For instance, simple switches could be used, since they have just two states – on and off. Storing numbers in base 2 is exactly the same as in base 10 – for instance, the number 1101 in binary is 13 in decimal:

$$1101 = (1.2^3) + (1.2^4) + (0.2^1) + (1.2^0)$$

$$= 8 + 4 + 0 + 1$$

$$= 13$$

Now, we can consider how we would convert 13 to binary. There are two methods – one simpler to understand, but one simpler to implement in a program. Here is the first. We know that in binary 13 will be made up by adding some or all of the numbers 1, 2, 4 and 8. We would not need the next power of 2 – 16 – because this is bigger than 13. We will need 8 because $1 + 2 + 4$ is 7, which is smaller than 13. So, 13 will be made up of 8 (2^3) plus 5. Now, we need to work out what 5 will be in binary. We do this by the same process – we will need to use 4 because $1 + 2$ is not enough, so 5 will be made up of 4 plus 1. In this case, we are finished because 1 is 2^0. So, 13 will be 8 plus 4 plus 1 or in full

$$13 = 1.2^3 + 1.2^2 + 0.2^1 + 1.2^0$$

$$13 = 1101$$

We can take what we have done and turn it into an algorithm. To convert N from base 10 to base 2

1. Find the largest power of 2, which is smaller than N – call it P1.
2. Note which power of 2 it is – this is the leftmost digit of the binary number.
3. Subtract P1 from N to give a new value of N.
4. Repeat steps 1 through 3 until N is zero at step 3.
5. The powers of 2 – P1, P2 and so on – are the digits of the binary number. Any power of 2 not used is represented by a zero.

An alternative method which is easier to produce an algorithm for is to keep dividing N by 2, noting whether the remainder is 0 or 1 until N becomes 0. The division by 2 is done an integer division so that, for example, 3 divided by 2 is not 1.5, but 1 with a remainder of 1. Here, is a table of how this works in the case of 13:

Number	Number divided by 2	Remainder
13	6	1
6	3	0
3	1	1
1	0	1
0	0	0

The first remainder is 1 – this is the remainder when 13 is divided by 2. It is also the rightmost digit in the binary number. One way of thinking about this is that odd numbers will always have 1 as their rightmost digit, while even numbers will always end in a zero. When we divide by 2 the second time, we are effectively dividing by 4 so the remainder will be our next binary digit.

The last row of the table shows that when the number becomes zero, the process will simply carry on producing remainders of zero. This means it is simple to produce a program which will give the binary to any desired number of bits – an 8-bit byte or a 32-bit word, for instance.

CONVERSION BETWEEN BASE 2 AND BASE 4

Conversions between base 2 and base 4 are much simpler than the general case because of the fact that four is simply two squared. This means that every base 4 digit is represented by exactly two base 2 digits as follows:

Base 4	Base 2
3	11
2	10
1	01
0	00

To convert a number from base 2 to base 4, all you have to do is group the base 2 digits in pairs, starting at the right, and replace each pair with the equivalent base 4 digit. So, to convert 11011_2 to base 4:

$$11011_2 = 01\ 10\ 11 = 123_4$$

Note that the number has to be padded out to an even number of binary digits on the left, so the leftmost 1 becomes 01. Conversion from base 4 to base 2 is simply the reverse process – replace each base 4 digit with the appropriate pair of base 2 digits.

The same principle applies to conversion between any number bases, which are powers of 2. One that is frequently used is the conversion of binary numbers to base 16, or hexadecimal. Since base 16 has more than nine digits, letters A–F are used for the extra digits. So, the conversion table between binary and hexadecimal is as follows:

Decimal	Binary	Hexadecimal
0	0000	0
1	0001	1
2	0010	2
3	0011	3
4	0100	4
5	0101	5
6	0110	6
7	0111	7
8	1000	8
9	1001	9
10	1010	A
11	1011	B
12	1100	C
13	1101	D
14	1110	E
15	1111	F

The advantage of using hexadecimal is that it provides a convenient way of representing long binary numbers. For instance, a 32-bit word can be represented using eight hexadecimal digits instead of 32 binary digits. Error messages which report memory addresses or values commonly use hexadecimal for this reason.

8.3 Raster Algorithms: Area Calculations

The second algorithm we will consider is one which we considered in the context of vector GIS in Chapter 5 – calculating the size of an area. In vector, the procedure is quite complicated because areas are defined as complex geometrical shapes. In principle, the raster method is relatively straightforward. To find the area of the feature labelled A (shaded area in Figure 8.6), we simply count up the number of A pixels. As we shall see things are more complicated when we use a quadtree, but let us start with our simplest data structure – the array.

The algorithm to calculate the size of the shaded area A is pretty much the same as the above description:

```
Array IMAGE[0..maxcol,0..maxrow]
area = 0
for i = 0 to maxrow
for j = 0 to maxcol
        if IMAGE[i,j] == 'A' then area = area + pixelsize
```

The program loops through each column in each row increasing the value of the variable area by the size of one pixel each time an A pixel are found.

The program for the run length encoded data structure is actually not very different. The data structure is shown again in Figure 8.7. Each run already

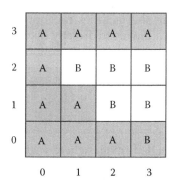

FIGURE 8.6
Example raster layer.

Address	200	201	202	203	204	205	206	207	208	209	210	211	212	213	214	215
Value	3	A	1	B	2	A	2	B	1	A	3	B	4	A		

FIGURE 8.7
Storage of image in run length encoded form in computer memory.

contains a count of the number of pixels, so for each run of type A, we simply use this to update our area variable.

```
area = 0
for i = 1 to nruns
if run.type == 'A' then
        area = area + pixelsize*run.length
```

It is easy to see that the area calculation algorithm has $O(n^2)$ complexity using a two-dimensional array, but $O(n)$ complexity with a run length encoded image (assuming that the number of runs is linearly related to the layer size). In this case, the run length encoded data structure can actually process this query more efficiently than the simple array, in contrast to the pixel value query.

Before moving on to look at the quadtree algorithm, it is worth looking at the algorithm for the run length encoded structure to illustrate a number of points about the way in which algorithms are actually implemented in practice. To implement this algorithm, we would need some way of storing and accessing the run length encoded image. One method is to use a one-dimensional array, and steps through these two elements at a time:

```
Array RLE[0..nruns*2]
for i = 1 to nruns step 2
if RLE[i] == 'A' then
        area = area + pixelsize*RLE[i + 1]
```

Some languages contain the facilities for more complex data structures than an array, and would allow us to directly declare a structure which contained two elements. For instance, in C, the following could be used:

```
struct rle {
integer: runlength;
integer: runtype}
rle image(5);
```

This sets up an array called image where each element of the array is a pair of items – the run length and the pixel value associated with it. Using this structure, something like our original pseudocode could be used in which we loop through each run, referring to both its length and type as required.

(a)

Address	200	201	202	203	204	205	206	207	208	209	210
Value	5	6	A	2	B	1	A	3	B	4	A

(b)

Address	200	201	202	203	204	205	206	207	208	209	210
Value	6	A	2	B	1	A	3	B	4	A	−1

FIGURE 8.8
Two techniques for keeping track of the number of runs. (a) Count at the start of the data. (b) Negative 'flag' value to signal the end of the data.

The pseudocode assumed that we knew how many runs there were in the image – but how would this information actually be stored? In practice, this could be handled in a number of ways. One would be to alter the data structure, to put a count of the number of runs at the start, as shown in Figure 8.8a. The program would then read this in as the value of n runs. The second, which avoids having to count up the number of runs, is to put what is sometimes called a 'flag' at the end of the data – in Figure 8.8b, this is a negative run length – which signals the end of the data.

A final point is that this algorithm would be more likely to be programmed as shown below:

```
area = 0
npixels = 0
for i = 1 to nruns
if run.type == 'A' then
        npixels = npixels + run.length
area = npixels*pixelsize
```

The difference is that rather than keeping a tally of the area, as we did for the full array, we keep a tally of the number of pixels (line 5) and then multiply this by the pixel size at the end (line 6). The reason is that addition is a much quicker operation on the computer than multiplication. In this second version of the program, we perform as many additions as there are runs, but only one multiplication – the first version performed an addition and a multiplication for every run. In overall terms, both algorithms have O(n) complexity, where n is the size of the raster layer. However, we can also consider the processing efficiency of the details of how the algorithm is coded. Even if we assume that multiplication and addition run at the same speed, then the first algorithm will need $2n$ operations, where n is the number of runs in the image, while the second will need $n + 1$. If we are dealing with an image with several hundred runs, then a change that at least halves the running time is worth having. This simple example illustrates that the analysis of efficiency can be applied to all levels of algorithm design, from the overall approach down to the slightest detail.

Address	200	201	202	203	204	205	206	207	208	209	210	211	212	213	214	215
Morton code	0	1	2	3	10	11	12	13	20	21	22	23	30	31	32	33
Value	A	4	8	12	A	A	B	A	A	B	B	B	B	A	B	A

FIGURE 8.9
Storage of quadtree in memory.

Let us now consider how we perform the area query using the quadtree data structure. Counting pixels will no longer work since we have replaced the original pixels with new ones of different sizes. If we look at the data structure itself (Figure 8.9), we can see that it would be an easy matter to go through counting the cells labelled A, but there is nothing which explicitly tells us how large each A is. This information could be stored of course, but it would increase the size of the quadtree, since each group of four elements in the structure would have to have a code indicating how large any leaf nodes in that particular group of four were.

A solution which avoids having to store the information is to start at the top of the tree, and visit each leaf in turn, keeping track of what size it is. To explain the principle, Figure 8.10 illustrates the quadtree shown in Figure 8.9, but in diagrammatic tree format, and showing the values at the leaf nodes.

We start at the root node of the tree. In Figure 8.10, this is actually the node at the top of the diagram; but, since it represents the original image, which is then subdivided, it is known as the root of the quadtree. We know the area of the root node, since it is the area of the original layer – 16 (assuming for simplicity that each pixel has an area of 1). We visit the first node in the tree, which is a leaf node with a value of *A*. We have come down one level so the node area is now 16/4 = 4. We therefore add 4 to our tally of area and continue. We continue by going back up the tree, and coming down to the next node. The area is again 16/4, but this time we have reached a branch. We therefore go down the first node from this branch, which will have an area of 4/4 = 1. This is a leaf, which has a value of *A*, so we add 1 to the area tally. We then go back up, and down the next branch – this is a leaf, but has a value of *B* so we ignore it and carry on. We continue this process, until we have traversed every branch of the tree in sequence, by which time we will have our answer.

This is fine diagrammatically, but how would we program this algorithm in our computer? Let us start by considering the four nodes below node 1 – these have Morton addresses 10, 11, 12 and 13 and contain the values ABBB (Figure 8.10). To add up the area of A pixels for these four leaves, we could have a program which looked as follows:

```
leafarea = 1
for node = 0 to 3
if value == 'A'  area = area + leafarea
```

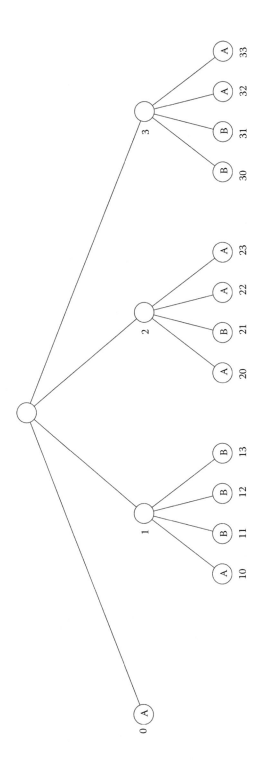

FIGURE 8.10
Quadtree representation raster image from Figure 8.9.

In other words – visit each leaf in turn and if its value is 'A' add the value of leafarea to the total. Now consider the four branches coming down form the root of the tree. These can be dealt with using the same approach, but with one difference – some of these nodes are not leaves but branches, so we will need to deal with them differently. Let us first modify the program to include the test for whether a node is a leaf or a branch:

```
leafarea = 4
for node = 0 to 3
if node is a LEAF then
        if value == 'A'  area = area + leafsize
else /* node is a branch
        Count 'A' leaves under branch
```

If we find a node is actually a branch, then we will need to have a program which will visit the four nodes under this branch, checking whether they are leaves with value *A* and if so, adding up their area. But, this is exactly what the program we are currently writing does! Does this mean that we have to write a second program, exactly the same as this, which can be called for nodes at the second level of the tree? Fortunately, the answer is no – all we do is call the program from within itself, as shown in the full algorithm below.

```
procedure area(leafsize)
local leafarea = leafsize/4
for node = 0 to 3
if node is a LEAF then
        if value eq 'A'  area = area + leafarea
else
        call area(leafarea)
```

We have now given our program a name – area. When it is called, the value of the current leaf size is passed as a parameter called leafsize. The program then sets up a local variable, called leafarea, which it sets to be the new size of leaves – a quarter of the previous value. The program then visits each of the current four nodes in turn, looking for leaves labelled A. If it finds another branch node, it calls itself again passing it the value of leafarea. The newly called version will receive this value and set up its own local variable to store the size of leaves at the next level down – although this is also called leafarea, it will be kept quite separate from the first leafarea variable. To complete this algorithm, we also need a program which will set the initial value of leafarea at the root of the tree, and call area for the first time:

```
program quadtree_area
layersize = 16
call area(layersize)
```

When a computer program calls itself in this way, it is termed recursion. As we can see, in this case, it has provided a very elegant way of traversing our quadtree structure.

Further Reading

As with vector algorithms, there is a good deal in common between the algorithms for raster GIS and for raster computer graphics. Indeed, the common ground is much broader in this case because similar methods have application in remote sensing, image processing, machine vision, pattern recognition and two-dimensional signal processing. There is even some commonality with computer modelling techniques such as genetic algorithms and cellular automata, which also use a grid of cells as their basic data structure. A good general text on image processing is Rosenfeld and Kak (1982). Of the remote sensing texts, Mather (1999) contains a certain amount on the operation of standard functions such as filtering and image rectification. The standard computer graphics text, Foley et al. (1990) covers raster as well as vector operations.

In the chapter, some simple algorithms were used to demonstrate the idea that raster data in quadtree format could be used to answer queries directly, without first converting the data back to a simple, raster array. Some further simple examples are covered in the NCGIA Core Curriculum Unit 37 (http://www.geog.ubc.ca/courses/klink/gis.notes/ncgia/toc.html). The two books by Samet (1990a, 1990b) cover a much wider range of algorithms, such as raster overlay, and illustrate how these too can be implemented using quadtrees.

An area where GIS and image processing come together is in capturing data from scanned maps using image processing methods. De Simone (1986) was one of the first to use automated image processing algorithms to identify vector features on scanned maps. More success has been achieved using semiautomated methods, which combine the image processing abilities of the human eye and brain with tools such as automatic line followers. Devereux and Mayo (1992) made some important developments in this area in the case of capturing vector data, and the approach has been extended to raster data capture by Wise (1995, 1999). Wise (1995) also discusses some of the programming issues which arise when handling raster data on a machine where memory is a limitation.

Something that has not been explicitly covered here is the conversion between vector and raster. The two articles by Peuquet (1981a, 1981b) discuss some of the early thinking on this, when it was an issue which dominated a good deal of the GIS literature. Jones (1997), in Chapter 8, provides a summary of some of the main algorithms, including Bresenham's algorithm for rasterising vector data. This is a particularly important operation in computer graphics, since computer screens are raster devices, so drawing any type of vector linework requires a conversion to raster. Bresenham's algorithm is important because, like the example of handling Morton codes in the chapter, it takes advantage of the greater speed of integer arithmetic, especially when this involves powers of 2.

9

Data Structures for Surfaces

Chapters 1 through 8 of this book have introduced the main issues involved in the handling of vector and raster data on the computer. Many of the main data structures for handling these two forms of spatial have been described as well as a selection of algorithms for carrying out some fundamental operations. In Chapters 9 through 11, we will consider two types of spatial data which can be represented using either vector or raster data. In the case of surfaces, we will see that raster data structures and algorithms have some advantages over vector, and in the case of networks, the opposite is true.

So far in this book, we have concentrated solely on data structures and algorithms that are suitable for two-dimensional (2D) data. However, the third dimension is extremely important in many types of spatial data, and handling this extra dimension has required the development of specific data structures and algorithms. In the real world of course, objects exist in three-dimensional (3D) space, but for many purposes it is possible to ignore the third dimension, and produce a model which is still a useful approximation of reality. Maps in general are a good example of 2D models of 3D space, but topographic maps also show one aspect of the third dimension – the elevation of different points. On most modern map series, the representation is in the form of contours, which as well as allowing an estimation of the height of points also portray general characteristics of the surface such as the steepness and direction of slopes.

Contours are not a true 3D representation. They can only be used to represent the upper surface of the landscape, and cannot show the existence of overhangs, or the shape of true 3D objects, such as geological strata. However, the surface itself plays an extremely important role in many natural and human processes. It is a key control in the distribution of both heat energy and water across the landscape, and thus is a key factor in the spatial pattern of climate, soils and vegetation. It is also a key control on the distribution of human activities, and an important factor in the location of settlements, businesses and transport routes. For these reasons, data structures have been developed for representing the land surface in a GIS, and algorithms developed for displaying and analysing them.

9.1 Data Models for Surfaces

That fact that we are trying to represent a surface has two important consequences. First, as elevation varies continuously, it is not possible to measure it everywhere. All measurements of elevation, no matter how closely spaced, are samples. Second, what is of interest are not the properties of the sample points themselves, but of the surface, and this means we have to form a model of the surface in the computer. In storing these sample data inside the GIS, it is not enough to simply record the location and value as might be the case if the values related to discrete objects such as lamp posts or houses. The values must be stored in such a way that it is possible to use them to derive useful information about the properties of the surface, such as the height, slope and aspect at any point. To see what this means consider Figure 9.1, which shows the same set of points being used to represent two different geographical phenomena.

In Figure 9.1a, the points represent the location of customers for a store, whereas in Figure 9.1b, they are samples of height from a ridge, shown schematically using contours. With the customer database, most queries can treat each point individually. For instance, to find out how many customers live within a 5-km radius of the store, the X and Y coordinates of the store can be used with the X and Y coordinates of each customer in turn to calculate the crow fly distance between the two. This can be done for each point in turn, and all we need to know is the location of the customers and of the store. Compare this with what appears to be a very similar query using the point samples from the surface in Figure 9.1b – finding out the area of land which lies above 120 m (or to be more precise – the area as projected onto a horizontal

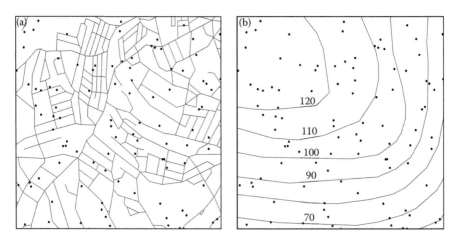

FIGURE 9.1
Points representing (a) customer locations and (b) spot heights. (© Crown Copyright Ordnance Survey, all rights reserved.)

surface, as opposed to the true surface area). If we simply select all points with elevations above 120 (Figure 9.2), this will not answer the query, for two reasons. First, to estimate the area, we need to be able to work out how much land lies between the sample points we have selected. This means that we need to know where the points are in relation to one another, so that we can join them up. If we do this, then as Figure 9.2 shows, we hit the second problem, which is that our sample points do not define the edge of the area we are interested in. What we actually need is to calculate the area above the 120 m contour, not the area between those sample points we have, which happen to be above 120 m. This means we need to use our sample points to estimate where the 120 m contour runs, and then use this line to answer the query.

In other words as well as storing the location and elevation of each point, it is also necessary to store information about the relationship between the points. There are three main models which have been developed which can store this additional information. The commonest is the grid model, in which the original sample heights are used to estimate the height at a series of points on a regular grid as shown in Figure 9.3.

In the case of the grid model, the relationship between the points is very straightforward. Every point has eight immediate neighbours (except for points along the edges of the grid) and the distances between them are either the grid resolution or, in the case of diagonal neighbours, this value times the square root of two. This means that there is no need to store this information explicitly, and the data can be stored as a standard raster grid. Using the grid model, we can provide an approximate answer to the query about the area of land above 120 m simply by selecting all grid points with a value above 120. Since the points are on a regular grid, we can assume that the height values

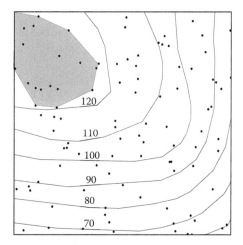

FIGURE 9.2
Problems with answering surface queries using point data. (© Crown Copyright Ordnance Survey, all rights reserved.)

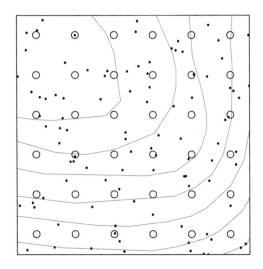

FIGURE 9.3
Grid data model for surface data represented by the open circles. The lines are contours and the dark points are spot heights. Either of these can be the source data for creating the grid. (© Crown Copyright Ordnance Survey, all rights reserved.)

represents the 'average' height for the square surrounding each point, so we simply multiply the number of points selected by the square of the grid spacing. Assuming the grid points are relatively closely spaced, this will give an answer which is approximately correct. A more accurate answer could be obtained by interpolating the 120 m contours line between the grid points.

The second model is the 'triangulated irregular network' or TIN. This takes the original sample points and connects them into a series of triangles, as shown in Figure 9.4 (the way in which the points are connected into triangles is described in Section 9.3). The relationship between the points is now stored using this triangular mesh.

To answer the query about land above 120 m, one possibility would be to select triangles with at least one corner above 120 m. For those with all three corners above 120 m, the area of the triangle can be calculated very simply. Where a triangle has one or more corners below 120 m, as with the shaded triangle in Figure 9.5, then some estimate would have to be made of how much of the triangle's area fell below 120 m. This triangle has two corners below 120 m. Inspecting each side of the triangle in turn, it is easy to discover which sides cross the 120 m contour and since the height at each end is known, it is possible to estimate how far along the side the 120 m contour line will cross. The crossing points on the two edges can then be joined to produce a quadrilateral, whose area can be calculated. This will give reasonable results as long as the triangles can be assumed to have flat surfaces – where the terrain is highly curved, or where the triangles are large, a more sophisticated approach would be needed. However, what this simple

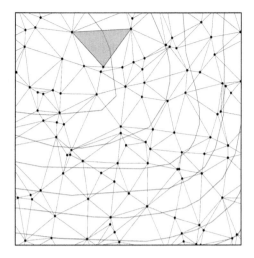

FIGURE 9.4
TIN model for surface data. Triangles have been created by joining the spot heights. (© Crown Copyright Ordnance Survey, all rights reserved.)

example illustrates is the way in which the connection between the points is the key to answering the query.

The final model takes contours as its starting point. Contours are not only useful as a graphical means of displaying the nature of the surface but also equipotential lines for the potential energy possessed by water flowing across the landscape. As water flows down the steepest slope, it will always

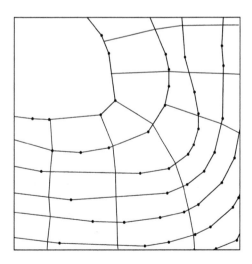

FIGURE 9.5
Contour-based model for surface data. (© Crown Copyright Ordnance Survey, all rights reserved.)

cross a contour line at 90°. Therefore, if a series of lines is drawn down the slopes, normal to the contours, then these, plus the original contours, define a series of 'patches' which define the surface as shown in Figure 9.5 (Note: for simplicity, these lines are shown as approximately straight – in fact, they will be curves, unless the slope is completely planar between the two contours).

This is the most complex of the data models for surfaces because the basic elements used to define the surface are no longer of uniform shape as they were in the case of the grid model and TIN. Information has to be stored, not only about the height at the corners of the patches but also about the shape of the edges and how the height varies across the patch. The models to do this have been developed for representing solid objects in computer-aided design software, since the method is well suited to representing smoothly curved objects such as car bodies or engineering parts. However, it is not available in standard GIS software packages, and is only used by a small number of researchers, so it will not be considered any further in this book. However, it is the model which would most easily give an accurate estimate of the area of land above 120 m, since this is simply a matter of selecting all patches where both bounding contours have a height of at least 120 m, and summing their areas.

This brief summary has indicated that there are two distinct sets of algorithms in the case of surface models in GIS. The first are the methods used to construct the model from the original data – to estimate the height at the grid points for the gridded model, and to construct the triangulation for the TIN model. The second are the methods used to answer queries using the constructed surfaces. A key issue in both cases is that of error. Since the surface can never be sampled or represented completely, the results of any analysis will always be estimates and will contain some element of estimation error. Keeping this to a minimum is one of the key issues in the design of algorithms for handling surfaces.

9.2 Algorithms for Creating Grid Surface Models

Figure 9.6 shows an area near the village of Slapton in Devon in the United Kingdom using three gridded models of the surface.

First, a brief note on terminology. There is a good deal of confusion in the GIS literature about the terminology for models of surfaces. In this book, the term 'digital terrain model' (DTM) will be used to refer to all models of the surface, whatever their form. However, it is worth noting that some authors use the term DTM to refer to the gridded model. To avoid confusion, when particular types of DTM are meant, such as the gridded DEM or a TIN, this will be made clear.

The ASTER GDEM was produced by a process called interferometry from satellite data, the Ordnance Survey PROFILE DTM was produced

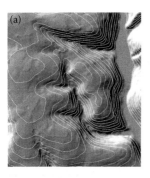 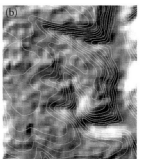 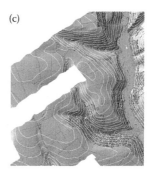

FIGURE 9.6
Elevation as represented by (a) 30 m ASTER GDEM, (b) 10 m Ordnance Survey PROFILE (© Crown Copyright/database right 2013. An Ordnance Survey/EDINA supplied service) and (c) 1 m Channel and Coastal Observatory LiDAR. The LiDAR data are for studying the coast, which is why they do not cover the whole catchment.

using contours as the data source, and the LiDAR dataset was produced from readings taken from a laser scanner mounted on an aeroplane. (In fact, the flight lines are visible running roughly SW-NE across Figure 9.6c). The ASTER and LiDAR DTMs are both in fact digital surface models as they represent the height of the upper surface which is visible from above, and this includes trees and buildings, whereas the PROFILE DTM is a bare-earth DTM.

For all three DTMs in Figure 9.6, the original data were not collected on a regular grid and so a process of interpolation was necessary to create the estimates of elevation in each of the grid cells from the source data. An entire book could be written about the ways in which DTMs can be interpolated and so it is not possible to cover the whole range in this chapter. To illustrate the issues which can arise when interpolating, the focus will be on interpolation from contours and points because the methods are relatively simple to explain. The techniques for generating elevation values from satellite remote sensing and aerial photography are completely different but in assessing the resulting DTM, the same two questions can be asked:

1. How accurate are the heights?
2. How well does the DTM model the surface?

The first question is the obvious one to ask and DTM producers usually publish statistics on the accuracy of their products, measured as the root mean square error (RMSE), which is calculated as

$$\text{RMSE} = \sqrt{\frac{\sum (z_e - z_o)^2}{n}}$$

where
z_e = estimated height from the DTM
z_o = observed height from data of higher accuracy
n = number of test points

The values for the three DTMs in Figure 9.6 are of the order of 20, 2.5 and 0.2 m for ASTER, PROFILE and LiDAR, respectively, although in all cases this varies depending on the nature of the terrain. The second question of how well the DTM models the surface is equally important but is rather more difficult to assess numerically. The three DTMs in Figure 9.6 all capture the essential shape of the valley of the Slapton Wood stream in Devon, UK. However, all of them also show some artefacts as a result of the way they were created. For instance, the ASTER GDEM has a 'bumpy' appearance, partly due to the fact that it senses vegetation but also due to errors in the estimation of the height. The Ordnance Survey PROFILE DTM has features which look like banks on some of the ridges, but which are small, systematic errors in the data. LiDAR has the fewest artefacts, but does contain one 'hole' where there are no estimates of height. The height errors produced by these artefacts are small, but they can result in systematic errors when the DTMs are used in GIS analysis as we shall see. This means that when considering interpolation algorithms, efficiency is not the main issue as it has been with the other algorithms considered in this book, but the quality of the resulting DTM.

In creating a gridded DEM, the task is to use the sampled elevation data such as the points or the contours shown in Figure 9.7a to estimate the

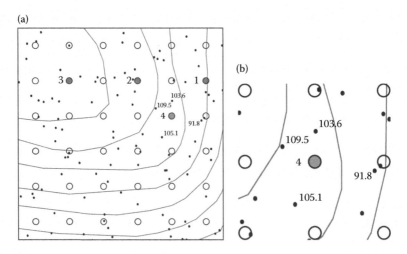

FIGURE 9.7
Small ridge showing sampled elevations (dark spots), contours and DEM grid points (grey circles). The shaded circles are referred to in the text. (© Crown Copyright Ordnance Survey, all rights reserved.)

elevation at a series of points spaced on a regular lattice. The algorithm only needs to be able to interpolate height at a single point at a time, since we can just apply it repeatedly to produce the complete grid. In thinking about how to design a point interpolation algorithm, it is instructive to think about how we might do the same task ourselves. First imagine that we have the contours shown in Figure 9.7, and we need to estimate the height at point 1. This point lies just above the 90 m contour and so its height will be very close to 90 m. Measuring the distance between the 90 and 100 m contours at this point, we find that point 1 is about 5% of the way between them, and so its height would be around 90.5 m. Point 2 is about 40% of the distance between the 110 and 120 m contours, so we could estimate its height at 114 m. However, the slope shape here is convex, with flatter slopes above point 2 and steeper below. This means that its height might be a little higher than the distance between the contours would suggest, at perhaps 114.5. Point 3 is more difficult. It is clearly about 120 m but the problem is that we do not know how far above. The ridge top appears to be quite flat so we might estimate it at 122 m, but with the available evidence this is a fairly rough estimate.

This simple example illustrates two important points about this process. First, in estimating the height at an unknown point, we use information from nearby rather than far away. Second, in doing the interpolation, we also use information about the shape of the surface in addition to the height data. Almost all computer algorithms for point interpolation use the first of these strategies, but differ in the way that they decide which points to use. However, they do not all use the second strategy, and this can lead to some major differences in the results they produce. As an example, let us consider one of the commonest algorithms for point interpolation, commonly called inverse distance weighting.

Consider DEM grid point labelled 4 in Figure 9.7. What value should we assign to the grid point? It seems clear that the value is likely to be similar to those for the sample points surrounding it, so one way of estimating the value is simply to take an average of the nearby data points. The selection of which points to use is an important issue which is discussed in detail below. To illustrate the principle of the method, let us use the four nearest points which are shown in more detail in Figure 9.7b:

$$z = \frac{109.5 + 103.6 + 105.1 + 91.8}{4} = 102.5$$

This value seems too low given how far the point is from the 100 m contour. The problem is that equal weight is being given to the point with height 91.8, which is roughly twice as far away as the other points, What we have to do is to give greater weight to the sample points which are near our grid point than those which are far away. This means that we need to measure the distance between the grid point and each of the sample points we use, and then use this distance to determine how important that sample point's value will be. There are several functions we can use to do this, and two of the commonest are shown in Figure 9.8.

The equations for these functions are as follows:

$$\text{Exponential } w = e^{-d}$$

$$\text{Power } w = d^{-1}$$

In both cases, d is the distance, and w is the weight and the functions produce weights which decrease towards zero as the distance increases. To apply this to the above example, we would calculate the distance to each point in turn, use this to calculate a weight using one of the formulas, and then calculate a weighted average. For example, applying the power function to the previous example:

$$z = \frac{0.042 * 109.5 + 0.050 * 103.6 + 0.031 * 105.1 + 0.025 * 91.8}{0.148} = 103.6$$

Notice that we no longer divide by the number of points (4), but by the sum of the weights. We now have an answer which seems more reasonable because the weight for 91.8, at 0.025, is smaller than the weights for the nearby points. It is also possible to change the relationship between the relative weights and the distance by adding an extra parameter into the equation for calculating the weights. This is easier to illustrate in the case of the exponential weighting function, which is changed to become

$$\text{Exponential } w = e^{-pd}$$

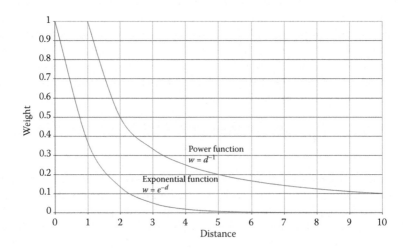

FIGURE 9.8
Two functions for assigning weights in point interpolation.

Figure 9.9 shows how the weighting function changes as p is changed. When p is 1, we get our original function. Making p greater than 1 reduces the weight given to nearby points, and making it smaller than 1 increases it. For instance, using $p = 0.5$ for the points in Figure 9.7 gives an estimate of 104.3 for grid point 4.

The difficulty is how to choose what value p should take. This is not the only problem. We also need to choose what sort of weighting function to use in the first place – two are shown in Figure 9.9, but which would be more appropriate for a given set of data? Both curves produce weights which decline asymptotically towards zero as d increases. (This means that they get closer to zero, but never actually equal zero until d equals infinity). However, they behave very differently for small values of d. When d is zero, the exponential function produces a weight of 1, but the power function returns a value of infinity. In practice, when d is zero, this means the point to be estimated is exactly coincident with one of the sample points, and so the z value of this is used as the interpolated value. However, for small values of d, the power function will produce much larger weights than the exponential function, meaning that it assigns much greater importance to nearby points. However, it may be difficult for the GIS user to know whether this is desirable or not.

There is also a further problem. In the simple example above, the nearest four points were used to make the calculations, to illustrate the principle of the method, but in practice the number of points to use would also have to be chosen. Most packages offer a choice of taking either a fixed number of points or all points within a fixed distance, with default settings in both

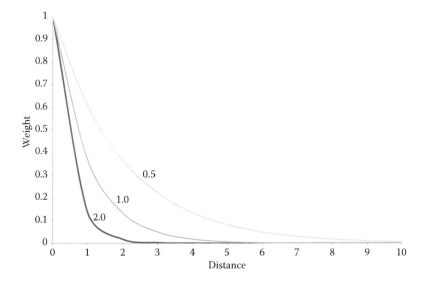

FIGURE 9.9
The effect of varying the exponent in the exponential decay curve.

cases. The problem is how to decide which option is best and which number or distance to choose.

One solution to some of these difficulties is to use a method of interpolation called kriging, which provides answers to these questions by analyzing the data itself. The first stage in kriging is to construct what is called a variogram, which is a graph that shows how similar the data values are for points at different distances apart. Points which are close together will tend to have similar height values, but this similarity will decrease as we pick points which are further apart. Eventually, we will reach a distance at which there appears to be little relationship between the data values. This distance can then be used as a cutoff point in selecting points to consider in the weighting procedure. What is more, the shape of the variogram can be used to determine what sort of weighting function to use. Kriging is a statistical technique, and can only be used if the data being interpolated satisfies various conditions. A full explanation of the method and its limitations is beyond the scope of this book, but some references are given in the reading section for those who want to know more.

Even when kriging is appropriate, it does not overcome one of the main drawbacks of the inverse distance weighting approach, which is that it does not attempt to use information about the shape of the surface. This problem becomes particularly apparent when the method is used with contour data rather than randomly scattered points as shown in Figure 9.10.

Figure 9.10 illustrates when a fixed distance is used to select the points from which to interpolate height. The two circles show the points which will be selected for the two grid points shown. The dark circle selects points from the 120 and 110 m contours, whereas the lighter circle selects points from the

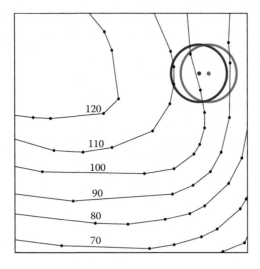

FIGURE 9.10
The use of inverse distance weighting with digitised contours. (© Crown Copyright Ordnance Survey, all rights reserved.)

110 and 100 m contours. This means that the interpolated height will change quite markedly between the centre of the two circles. The effect of this can be seen in Figure 9.11, which shows the DEM produced by applying the inverse distance weighting method to a set of contours. Around each contour, the interpolated heights are very similar to the contour height. Midway between each contour, there is a rapid change in the interpolated heights, leading to a DEM which resembles a 'wedding cake'.

These distinctive ridges are an example of what are known as interpolation artefacts. The inverse distance weighting parameters have been deliberately chosen to exaggerate the effect, but it will exist in a less pronounced form whatever parameter values are used. Remember that whenever we interpolate a value there will be some error in our estimation. This error will affect any further processing we do using the data, but as long as the amount of error is relatively small the effects should not be too great. However, in the case of DEM interpolation, the errors often have distinctive spatial patterns, and form artefacts. The problem is that although the errors in the estimated height may be small in absolute terms, they may cause quite large errors in our analysis. For instance, if we calculate slope angles using the DEM in Figure 9.11, we will get values which are consistently too low around the contours, with values which are too high in between. On average, the values may be about right, but if we use a map of the slope angles to identify areas suitable for ploughing or to estimate soil erosion rates, then our results will be strongly affected by the presence of the artefacts.

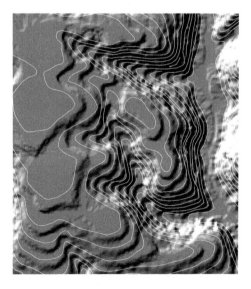

FIGURE 9.11
Gridded DEM produced using inverse distance weighting on contours. (© Crown Copyright Ordnance Survey, all rights reserved.)

This illustrates that in assessing GIS algorithms, speed and efficiency are not the only concerns, and we must also be conscious of the accuracy of the results which are produced. In the case of DEM generation, we would only expect to create a DEM once, but we might use it hundreds or thousands of times, so the speed of the creation algorithm is almost irrelevant compared with the accuracy of the results it produces. We will return to point interpolation in due course, but first let us look at algorithms for creating the other major surface data model, the TIN.

9.3 Algorithms for Creating a Triangulated Irregular Network

In trying to create a triangulation, we are trying to join together our sample points in such a way that they produce a reasonable representation of the surface. We will look more closely at the use of the TIN model in Chapter 10, but to understand some of the issues in creating the triangulation it is useful to know that the key to the use of the TIN is the fact that the height of the three corners is known. Given this information, it is possible to work out which way the triangle is facing, how steeply it is sloping and the height at any point in the triangle. In other words, the triangles do not simply represent the height of the sample points, they model the whole surface.

In the same way that we interpolated heights at unknown grid points by looking at nearby points, we should expect to produce a triangulation by joining nearby points rather than distant ones. Let us start by looking at an algorithm which attempts to produce a triangulation based on this distance measurement alone.

At first sight this seems like quite a simple algorithm to produce. Given a set of points, simply calculate the distance between all pairs of points, then join the closest two, followed by the second closest and so on.

In the case of the points in Figure 9.12, the table of distances might be as follows:

	a	b	c	d
a		1.5	1.2	3.3
b			2.4	3.1
c				3.2
c				

Only part of the table is filled in as the distance from a to b is the same as from b to a, and the distance from any point to itself is zero. It is a simple matter to use this table to produce a triangulation – the shortest distance is between a and c so these two points are joined first, followed by a and b and so on. After five stages, the triangulation will look like Figure 9.13a, but

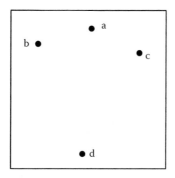

FIGURE 9.12
Example points to illustrate triangulation algorithms.

at this point a problem emerges. There is still one distance in our table, but joining a and d will produce the situation shown in Figure 9.13b in which the new triangle edge crosses one of the existing edges.

We cannot allow this to happen because we do not have a height for our crossing point (labelled e on the diagram) and so we would not be able to estimate any properties of the surface within the four triangles which have e as one of their corners. What happens if we do not force the two lines to intersect? We still have a set of triangles – abc, abd and acd – and we know the height at the three corners of all three. The problem is that parts of the surface are now defined by two triangles rather than one. For instance, the shaded area in the diagram belongs to triangles abd and abc. If we need to estimate the height of a point within this area, which triangle would we use? Clearly, then we would have to modify our distance-based algorithm so that it checks that a new edge does not cross any existing edges. Given what we already know about detecting whether lines cross, this will add significantly to the workload of the algorithm.

Obviously, our simple algorithm is going to be more complicated than we thought. We cannot simply keep joining points together, we need some way

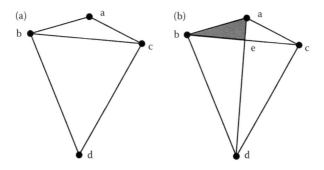

FIGURE 9.13
Triangulation after five (a) and six (b) stages.

of checking that we are producing a set of triangles. One possibility might be to join the closest pair of points, and then add two further lines to complete this triangle before doing anything else. We could then join the point which is closest to this triangle and complete our new triangle and carry on adding triangles to our set until we have used up all the points. This would work, but our resulting triangulation would be very dependent on where we started the process, that is where the two closest points were. A very small change in the location of one of the points could result in a different starting point for the process and a completely different set of triangles. This suggests our method cannot be producing a very good triangulation – if it was, then a small change in the position of the points should only have a small effect.

There is another weakness in this shortest distance approach. If we look at the triangulation on the left of Figure 9.13, we can see that the upper triangle is rather long and thin in comparison with the lower one. The best triangles to use for interpolation within our final TIN are 'fat' ones like the lower triangle. With thin triangles, there can be a number of problems. To estimate the height of a point on the surface for instance, we find out which triangle it falls in, then fit a flat surface through the triangle corners and estimate height using this. With a long thin triangle, any small errors in the location of our points or errors caused by the loss of precision will have a much greater effect on this process. We may locate the point in the wrong triangle, for instance. In addition, the calculation of the triangle slope will be far more sensitive to errors for thin triangles than for fat triangles.

Just to add to the woes of this method, it may have seemed simple but it is far from efficient. To begin with, we need to calculate and store all the inter-point distances. The number of distances in total is $n(n-1)/2$. Each point is compared with every other point except itself ($n(n-1)$) but since the distance from a to b is the same as b to a, we only need to do half the calculations. Therefore, both the computational and storage complexity are $O(n^2)$.

Creating the triangulation also has quadratic complexity. The majority of points in a triangulation lie on the junction of three triangles, so the number of lines is approximately $3n$ (but not exactly this, for reasons which are not important). If we have to check to see whether each line might intersect one we have already added, then for each of our $3n$ lines, we potentially have to check with up to $3n$ lines (fewer obviously in the early stages of the algorithm). However, we can see that this approach is going to have roughly quadratic efficiency in terms of both speed and storage.

All these suggest that simply joining nearby points together is not going to produce a very satisfactory result. The method which is actually used is based on the idea of producing triangles which are fat rather than thin, but it is easiest to explain this by starting with what is known at the Voronoi tessellation, which is shown in Figure 9.14.

Each area on this diagram represents the part of the surface that is nearer to the point it contains than to any other point. In fact, this tessellation can be used for the simplest form of interpolation of all, which is sometimes used for

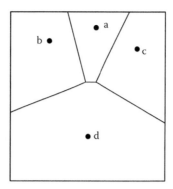

FIGURE 9.14
The Voronoi tessellation of the sample points.

estimating rainfall. If we assume that each point is a rain gauge, then for any other point on the surface, a simple way of estimating the rainfall is to take the value from the nearest gauge. This is often done when making rough estimates of rainfall over catchments when there are very few gauges available.

What is more interesting in our case is that Figure 9.14 is another example of what mathematicians call a graph, and as in the case of the street/block network we looked in Chapter 3, there is a dual to this graph as shown in Figure 9.15.

This graph is produced by joining together points in the Voronoi tessellation which share a common boundary. When the graph is represented using straight lines between the points, then what is produced is the Delaunay triangulation, after the mathematician who first studied its properties.

One of its most interesting properties is that it can be shown mathematically that this triangulation produces the 'fattest' possible triangles. What this means is that of all the different possible ways of triangulating a set of points, the Delaunay triangulation produces triangles in which the smallest angles within each triangle are as large as they can possibly be. Of course, long thin triangles will still be produced – when we look at the use of the

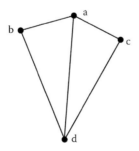

FIGURE 9.15
Delauney triangulation of the sample points.

Delaunay triangulation with some real data we will see this – but, on average there are fewer of them than with other triangulations.

Another useful property is that the triangulation is mathematically defined for almost any set of points (although it is possible to define sets of points for which the triangulation is not uniquely defined). It does not matter how we calculate it, it will always be the same (provided we do the calculations correctly!). What is more, if the points change slightly, the triangulation only changes slightly too, and only in the area near the altered point. Finally, it can also be proved mathematically that none of the Delaunay triangles edges cross each other, so there is no need to include a check for this in our algorithm.

Defining the triangulation is one thing – but how do we produce it in the computer? The answer lies in the use of some of its mathematical properties. Consider the two triangulations of the four points in Figure 9.16. The one on the right is clearly preferable to the one on the left, as both triangles have minimum angles which are larger than the minimum angle of the 'thin' upper triangle on the right. What is interesting is that this can be tested mathematically. Assume we have just added the edge bc to the triangulation and want to test whether it is a proper Delaunay edge (remember that the Delaunay triangulation is mathematically defined for a set of points – our task is to find it). Another property of the Delaunay triangulation is that a circle drawn through three points which are joined in the triangulation will not contain any other point. What we do is take one of the triangles that bc belongs to and draw the circle which goes through all three corners of the triangle. In Figure 9.16a, this circle which goes through b, c and a also include point d and so edge bc cannot belong to the Delaunay triangulation for this set of points.

However, if we do the same for the figure on the right, we will find that line bd does not cause this problem. Let us return to Figure 9.16a. The edge between b and c belongs to two triangles – abc and bcd – and together these

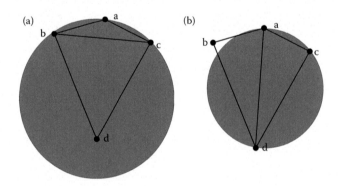

FIGURE 9.16
Edge flipping to produce the Delauney triangulation. (a) Joining b and c places d inside the circle, which breaks the rules of the triangulation. (b) Flipping this to join a and d restores the Delauney property.

form a four-sided figure. If we 'flip' edge bc to make it join the other two corners, then we will produce a legal Delaunay edge as shown on the right-hand side of Figure 9.16. Note that once we flip the edge, the definition of the triangles changes, and hence so does the circle, which now joins the three points acd but excludes point b.

This test and edge flipping will always work as long as all the other lines apart from the one we are testing belong to the Delaunay triangulation. This is what gives us the key to an algorithm for computing the full triangulation. Rather like the algorithm we used for sorting, the approach we take is to begin by triangulating just three points, and then adding extra points one at a time. In Figure 9.17a, points 1 through 3 are joined into a triangle and then point 4 is added. This is inside an existing triangle and so we create new edges by joining it to each of the three vertices. When point 5 is added (Figure 9.17b) it is outside the current set of triangles so it is joined to the two corners of the triangle nearest to it. However, this makes edge 2–3 illegal because a circle which passes through points 2, 3 and 4 will include point 5. Therefore, edge 2–3 is flipped so that we add edge 4–5 instead.

Adding a point inside an existing triangle can also produce the same effect. In Figure 9.17c, point 6 makes edge 4–5 illegal and so this is flipped and edge 2–6 added to the triangulation.

But how do we pick the three points to start off this process? In fact, it does not really matter. When we only have three points, the triangle that joins them is by definition the Delaunay triangulation of those three points. As we add each successive point, the edge flipping will ensure that we always have a Delaunay triangulation of the current set. However, as we proceed, some of the points which we add will be inside existing triangles, and some will be outside the triangulated area. To avoid having to distinguish between these two cases, we can start by producing a large triangle which completely encloses all the points. This means that every point we add has to be inside one of the existing triangles, simplifying the implementation of the edge testing procedure.

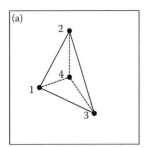 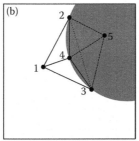 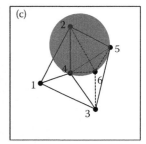

FIGURE 9.17
Incremental algorithm to build the Delauney triangulation. (a) Addition of first four points. (b) Edge flipping when a new point is added outside existing triangles. (c) Edge flipping when a new point is added inside existing triangles.

As there is no need to calculate the huge array of distances which we needed for the first algorithm, the storage requirements are much more modest, and in fact turn out to be O(n). The processing efficiency is O(n log n), which is the same as the sorting algorithm and for similar reasons – we need to process n points and for each we need to find out which of the current triangles it falls into. Now that we have discovered how to calculate the Delaunay triangulation, let us consider some of the issues in using it to create a model of the terrain surface.

9.4 Grid Creation Revisited

We have already seen that there are two desirable properties for algorithms which perform point interpolation – they should use information from nearby points, and they should attempt to model the surface. The methods described in Section 9.2 all satisfied the first condition, but did not really try to satisfy the second. In this section, we will look at some other algorithms that do try to tackle the second issue.

The first is the TIN model. At the heart of the construction of a TIN is the idea that the points which are joined are natural neighbours of one another. Hence, the Delaunay triangulation provides a different answer to the question of which points should be considered to be 'nearby'. It also provides a means of modelling the surface, as the triangulation produces a set of triangular patches which completely cover the area. Since we know the height at the corners of the triangles, we know which way they are facing, and so what we have is a model of the surface, as shown in Figure 9.18. This shows part of the valley which we saw in the previous section, modelled using a TIN rather than a gridded DEM. The contours have been used as the data

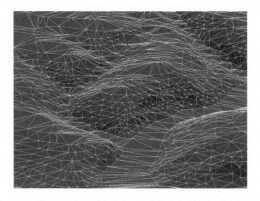

FIGURE 9.18
How a TIN models the surface. (© Crown Copyright Ordnance Survey, all rights reserved.)

source again, and the triangles which model the surface are clearly visible in this view.

To estimate height at an unknown point, all we have to do is find out which triangle it falls in and then estimate the height of the triangular patch at that point. This is relatively simple, but has a major drawback, in that the model is one in which flat triangles meet at sharp edges and these are clearly visible in the resulting DEM (Figure 9.20a). We have now got rid of the 'wedding cake artefacts', which a simple IDW approach created and produced something which is more realistic. The triangle edges are clearly visible; however, making the hillsides looks too faceted. However, the sharp break of slope where the hillsides meet the flat valley floor is quite a good model of the actual terrain.

A much more common way to model the shape of the terrain is to use a method based on polynomial curves. When using the TIN, the interpolation assumes that each TIN triangle has a planar surface. If instead we were to assume that the TIN surfaces were curved, it might be possible to model the change of slope between triangles more realistically. In fact, we do not need to use triangles at all because we can use polynomials to estimate height directly from the source data.

The principle behind this is easier to explain in 2D than in 3D. Figure 9.19 shows three polynomial curves, each containing successively higher powers of as follows:

$$\text{Linear}: z = 60x$$

$$\text{Quadratic}: z = 10x^2$$

$$\text{Cubic}: z = x^2 - x^3$$

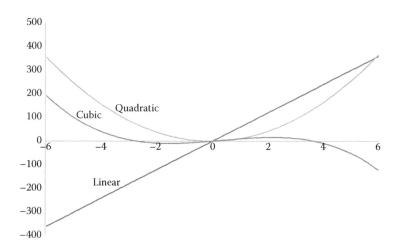

FIGURE 9.19
Graphs of polynomials of different order.

The linear function is a straight line, the quadratic function is a U-shaped curve, and the cubic function is like an S on its back. As the graph shows, each additional power of x effectively adds another 'bend' to the curve. The same principle applies to 2D polynomials, where the height of the curve (z) is a function of powers of the two horizontal dimensions, x and y, and the resulting curve forms a 2D surface rather than a one-dimensional line. The larger the powers of x and y, the more curved the surface will be. There is a large range of interpolation methods based upon fitting polynomial curves to elevation data and these tend to produce relatively smooth DEMs as shown in Figure 9.20b, which has been produced using a radial basis function.

The hillsides are now much smoother and do not have the unrealistic triangular artefacts which are visible in Figure 9.20a. However, the valley floor is no longer smooth but has a 'bumpy' appearance. This is because polynomial curves are smooth, and when the terrain changes shape very quickly, it is hard to make them fit. One way of thinking about it is to imagine that the contours in Figure 9.20 are made of stiff wire, and the task is to drape a sheet of plastic over them to create a model of the surface. If the plastic is too thin it will drape over the wires and sag in between them, so it needs to have some stiffness. However, if it is too stiff, it will be difficult to make it fit where the shape of the contours changes rapidly.

The truth is that there is no method of DEM interpolation which is universally best and the creation of DEMs usually means taking the time and trouble to experiment with different methods and with different parameter settings. This reinforces the point made earlier that in the case of algorithms for creating representations of surfaces the key issues tend to resolve around

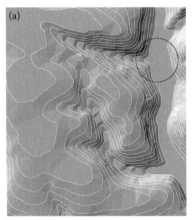 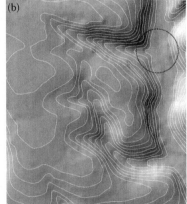

FIGURE 9.20
Gridded DEM produced using different interpolation methods. (a) Linear interpolation from a TIN. (b) A radial basis function. Notice the difference in the way the valley floor is modelled in the area indicated. (© Crown Copyright Ordnance Survey, all rights reserved.)

the quality of the final surface, rather than the efficiency of the algorithm. One of the problems, which is the subject of a good deal of current research, is that there is no simple, objective measure of this quality equivalent to the big-O notation for assessing the quality of different algorithms.

Further Reading

Units 38 and 39 of the NCGIA Core Curriculum describe the gridded and TIN models of surfaces, respectively (http://www.geog.ubc.ca/courses/klink/gis.notes/ncgia/toc.html). The vector/raster debate, which was so important in the early days of GIS development, has a counterpart in the debate about the relative merits of the gridded DEM and the TIN for representing surfaces. The TIN was developed partly as an answer to the perceived weaknesses of the gridded model, as explained in some of the early work by Peucker and colleagues (Peucker et al. 1978, Mark 1975, 1979). More recently, Kumler (1994) has undertaken a comprehensive comparative review, concluding that although the TIN seems intuitively a better representation of the landscape, many algorithms produce better results on a gridded DEM.

The use of contour-based models was pioneered by Ian Moore and coworkers at the Australian National University (Moore et al. 1988) and is well described in the book edited by two of the group (Wilson and Gallant 2000). The approach has never gained widespread usage but continues to attract interest because of its potential for modelling the surface (Mizukoshi and Aniya 2002). The description of the Delaunay triangulation algorithm in this chapter is partly based on the material in Chapter 9 of de Berg et al. (1997). Chapter 7 of the same book describes the Voronoi tessellation. The Voronoi tessellation and Delaunay triangulation both have uses outside surface interpolation in GIS. For instance, geographers know the polygons which form the Voronoi tesselation as Thiessen polygons, after a meteorologist who 'invented' them as a means of estimating areal rainfall amounts based on values from rain gauges. A good starting point to find further information on the range of applications is Chris Gold's Voronoi website (http://www.voronoi.com/). Both Gold (1992) and Sibson (1981) describe an interpolation algorithm based on the Voronoi tesselation. Interpolation has been the subject of a good deal of discussion. Unit 41 of the NCGIA Core Curriculum (http://www.geog.ubc.ca/courses/klink/gis.notes/ncgia/toc.html) gives a good overview of a range of methods. Burrough and McDonnell (1998), Franke (1982), Goodchild and Lam (1980) and Lam (1983) also review a wide range of different techniques. Burrough and McDonnell (1998) give a good description of kriging in particular and Franke (1982) includes a good description of radial basis function methods. A more recent review is available online at http://www.agt.bme.hu/public_e/funcint/funcint.html. Fisher and Tate (2006) provide a good introduction to the study of DTM error and Wise (2000b) discusses the limitations of RMSE as a measure of DTM quality. The creation of DEMs from contour data continues to attract interest and both Taud et al. (1999) and Hutchinson (1989) describe different methods which try to exploit the fact that contours contain useful information on the shape of the terrain. Toutin (2008) gives an overview of the ASTER project and further information can be found on the project website (http://asterweb.jpl.nasa.gov/). The other global elevation dataset was produced by NASA's SRTM mission (Farr et al. 2007, http://www2.jpl.nasa.gov/srtm/). There is a lot of interest in LiDAR data currently because of its fine resolution and high accuracy. The Open Topography website provides a good selection of reports on how LiDAR is processed (http://www.opentopography.org) as well as providing access to LiDAR data. The density of the original data points is so high in LiDAR that the creation of gridded DEMs is much less sensitive to the interpolation artefacts which are described in this chapter (Lloyd and Atkinson 2006) but LiDAR has its own sources of error as described by Leigh et al. (2009). In fact, the greatest challenge in using LiDAR effectively may be managing the large data volumes, as Chen (2007) discusses.

10

Algorithms for Surfaces

Now that we have seen how we can create the two main surface models, it is time to consider their use in answering queries about real-world surfaces. Again, it is important to remember that all our models are just that – models of the true surface. The degree to which the model is an accurate representation of the real surface will affect the degree to which the answers we get when using it reflect characteristics of the real surface.

We have already seen that many of the methods for interpolating a gridded DEM from elevation data give rise to artefacts which will clearly affect our results. In this chapter, we will see whether the results are also affected by the algorithms we use.

10.1 Elevation, Slope and Aspect

The most fundamental property of a surface is its height. If the point we are interested in happens to be one of the grid points in a gridded DEM, or one of the triangle corners in a TIN (i.e., one of the original sample locations), we can answer the query straightaway. In general, of course, this is unlikely to be the case, and we are going to need a method for estimating height at any point on the surface. We have already covered the point interpolation problem because this is how we created the gridded DEM from the original sample data. However, in that case, we did not have a surface model – simply a collection of points – and we can use the fact that we have a model to simplify the point interpolation process. In particular, both the DEM and the TIN tell us what the height is at nearby points on the surface, so that all we have to do is decide how best to use this information to estimate the height at our unknown point.

In the case of TIN, this is a two-stage process. First, we need to identify the triangle that our point falls into. Second, we use the information about this triangle, and possibly its neighbours, to estimate the height at the unknown point. Given a set of triangles and a point, we could use one of the point in polygon algorithms which were described in Chapter 5 to answer the first part of the query. However, these were designed to work with polygons of any shape and size. In this case, all our polygons are triangles, and all the polygons sides are made up of single straight lines and we can use this information to speed up our point in polygon algorithm.

Triangles are what are called convex polygons. Informally, this means that they do not have any indentations. Formally, it means that any two points within the polygon can be joined by a straight line which does not cross the boundary. Other examples of convex polygons are the regular geometrical shapes such as the square, rectangle and hexagon. This simplifies the point in polygon algorithm because a ray from a point will only cross the boundary once (if the point is inside) or twice (if the point is outside). In fact, rather than use the normal point in polygon test, we can exploit this property to produce a special purpose test.

If a point falls inside a triangle, then its *Y* coordinate must fall in between the *Y* coordinates of two of the triangle corners and likewise its *X* coordinate. This means we can use a very rapid test to exclude from consideration all triangles which lie entirely above, below or to one side of our point. In many cases, this will leave a very small number of triangles, which can be tested using the normal ray method.

Once we have found the correct triangle, we can find how steeply it is sloping and which way it is facing from the coordinate of the three corner points. The mathematics of how this is done is beyond the scope of this book, but an explanation of why this is always possible is useful, and has a bearing on some of the material which will be covered further on. The problem is to fit a –2D plane surface through three points in space. Let us start with the simpler problem of fitting a 1D line through two points in space. To simplify things even further, we will make one of the points the origin of our coordinate system (the point with coordinates 0,0) as shown in Figure 10.1.

The general equation for a straight line, which we used in Chapter 4, has the following form:

$$z = a + b \cdot x \tag{10.1}$$

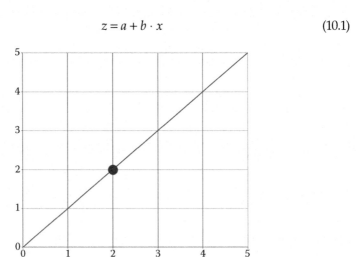

FIGURE 10.1
Graph of $z = ax$.

This has two unknown parameters – a and b. The value of b tells us how steeply the line slopes – for each increase in the value of X by 1, what will be the increase in the value of Z? The value of a is what is called the intercept – when X has a value of 0, what value does Z have? In this case, we know that when X is 0, Z is also 0, so in fact we only have one unknown quantity – b – and our equation will be

$$z = b \cdot x \tag{10.2}$$

We can see from looking at the graph that the value of b in this case is 1. When x increases by 1 unit, z also increases by 1 unit. But, if we had to work this out given the location of some points along the line, how many points would we need? The answer is just one. To work out the slope, we need to work out how far the line rises in Z for a given rise in X, and we can work this out from the position of one point on the line.

Now, consider the line in Figure 10.2. This no longer goes through the origin, and so the equation of this line is

$$z = b \cdot x + a \tag{10.3}$$

The line still has a slope of 1, but it is higher up the y axis – 1 unit higher up in fact, so b has a value of 1. This line has one more unknown parameter compared with the first one – a – and so we will need a second point along the line so that we can calculate this second parameter. Now, imagine that the line is extended to a third dimension, so that it becomes a flat surface in 3D space. If we only have our two known points, we are not going to be able

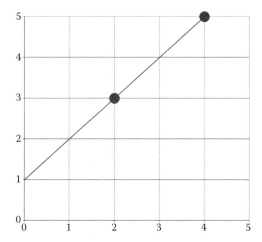

FIGURE 10.2
Graph of $z = ax + b$.

to tell exactly where this surface goes. Imagine taking a sheet of card and laying it along the straight line. We could spin the card through 360°, with it still lying along the line. However, as soon as we have a third fixed point, this will fix the card into position in space. The equation of the plane will be

$$z = a \cdot x + b \cdot y + c \qquad (10.4)$$

We now have three unknowns – a, b and c – and we have seen that we can work these out as long as we have three points on the plane. Since we have the height at the three corners of our TIN triangles, we can always fit a plane surface through them.

Once we have the equation of the plane passing through the triangle of the TIN, we can use the parameters to tell us the characteristics of the surface. Elevation is obtained simply by putting the x and y coordinates of our point into the equation. Slope and aspect can be calculated from the values of a and b, which tell us how steeply the plane is sloping in the X and Y directions, respectively. Note that a and b do not give us a slope in degrees, but as a proportion. If we look again at the line in Figure 10.2, the slope of this line can be calculated as follows:

$$\text{Slope} = \frac{z2 - z1}{x2 - x1} \qquad (10.5)$$

assuming the two points have coordinates $(x1,z1)$ and $(x2,z2)$. This calculation produces a value of 1 in this case, which is the tangent of the slope in degrees. The angle is therefore the arctangent of 1, which is 45°. Slopes calculated as the ratio between the fall in height divided by the horizontal distance in this way are sometimes multiplied by 100 and reported as a percentage. This is often seen on road signs warning motorists of steep gradients on roads.

Given values for a and b, the following formulae will calculate the maximum slope of the plane (also expressed as a proportion) and the direction it faces (as an angle with north as zero):

$$\text{Slope} = \sqrt{a^2 + b^2}$$
$$\text{Aspect} = \tan^{-1}\left(\frac{-a}{b}\right) \qquad (10.6)$$

This assumes that the TIN surfaces are modelled as planes. However, it may be more realistic to assume that the triangle surfaces are curved as we saw in Chapter 9. Many GIS packages use quintic polynomials (which contain terms up to powers of 5) for this purpose. Since these contain higher-order terms, we know that they can model surfaces which curve in several directions. In fact, using quintic polynomials, it is possible to ensure that the curves defining neighbouring triangle surfaces meet smoothly along the triangle edges, which removes the angular appearance of DEMs produced using plane triangles.

However, there will be more than three unknowns for the polynomial equations, since as well as parameters defining the slope in x and y, we also have parameters defining the amount of curvature in various directions. We cannot fit these surfaces using just three points, so we have to use points from the neighbouring triangles in addition. This makes intuitive sense. We are now modelling our triangle, not as a flat surface, but as a surface with a bulge in it. The only way to find out whether the triangle surface bulges a little or a lot is to look at the slope and direction of the neighbouring triangles – if these are very similar to the one we are interested in, then the surface must be fairly planar at this point, but if they are very different it must be quite curved.

Finding the elevation, slope and aspect at any point on a gridded DEM is the same in principle as with a TIN, but the details of the calculations are very different. The first stage is relatively trivial. With a TIN, we need a special algorithm to find out which triangle our query point lies in. With a grid, the points are spaced in a regular layout, so it is very easy to find out which grid cell the query point falls in – the coordinates of the point tell us directly. However, the cell itself will not help us answer our query as the triangle did. The cell is simply the square area immediately surrounding the grid point – it does not have a slope or direction as the triangle did. To estimate our surface parameters, we need to use the information from the neighbouring points. There are actually several ways in which this can be done.

The two equations for slope and aspect above are based on the slope in the x and y directions. In a grid, these are potentially very simple to estimate – for instance, in the grid in Figure 10.3, the slope in the y direction can be calculated from using the elevations of the points in the rows above and below the central point:

$$Y \text{ slope} = \frac{B - H}{2d} \tag{10.7}$$

$d =$ grid spacing.

A	B	C
D	•E	F
G	H	I

FIGURE 10.3
Location of cells in 3×3 window used for estimating surface properties at a point.

A similar calculation using the points from the neighbouring columns will give the slope in the x direction, and the two values can simply be inserted in the equations. By doing this, we are effectively assuming that the slope and aspect at the unknown point are the same as slope and aspect at the nearest grid point (i.e., the central point in Figure 10.3). Applying the same logic to elevation, we can assume that height at any point is the same as the height of the grid cell it falls into. There are several problems with this approach. One is that the results are very dependent on the accuracy of the heights in each of the grid points – a slight error in the height in pixel B, for example, will produce an error in the estimate for b. The same was true for the TIN method, as the elevation, slope and aspect were all essentially calculated from just three data points – the heights at the corners of the enclosing triangle. The difference was that these heights were measured, whereas the heights in a gridded DEM are all interpolated estimates, which means that in addition to any measurement error they will contain errors introduced by the interpolation process. For this reason, the equation used to estimate slope in the X direction at the central grid point is usually

$$Y \text{ slope} = \frac{(A + 2B + C) - (G + 2H + I)}{8d} \tag{10.8}$$

Thus, rather than simply looking at the difference between B and H, the difference between the average height of the points in the rows above and below is used.

Another method altogether is to fit a surface through the points in the neighbourhood, as was done with the TIN. This has the advantage that elevation, slope and aspect can be calculated at any point within the neighbourhood, not simply the central point. With the nine data points in a local window, such as in Figure 10.3, it is possible to fit a surface with nine parameters, and some algorithms do this. However, it may well be better to fit a somewhat simpler surface. To understand why, let us return to the example of fitting a straight line to a series of points.

Figure 10.4 shows a series of six data points which are estimates of elevation along an imaginary hillslope. It is possible to a fit a curve that will go through all six points exactly as shown by the solid line – this is a fifth-order polynomial. The graph shows that to pass exactly through each point, the curve has to bend quite sharply in place. If we used this curve to estimate height or slope at any point along the line, then some of our estimates will seem rather odd – for instance, towards the right-hand end of the graph, the hillslope will apparently have a negative slope, and appear to be facing in completely the wrong direction.

However, we can also fit a line which does not pass through every point exactly, but tries to capture the overall trend of the data points. In the case of the dotted line in Figure 10.4, a third-order polynomial has been used. We can see that at the upper end of the slope, the line traces a more plausible

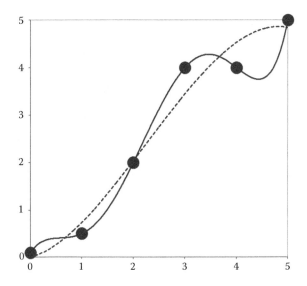

FIGURE 10.4
Lines through a series of point. Solid line – fifth-order polynomial. Dotted line – third-order polynomial.

shape between the actual data points. This line will also be much less sensitive to small changes in the data points. The same principle applies to fitting a surface through the points in our 3×3 window in Figure 10.3. A surface which goes through all nine data points will be far more sensitive to errors in the DEM than a simpler curve.

As an illustration of the difference which algorithms can make to the results of using a DEM, consider the three maps in Figure 10.5. All three are maps of aspect derived using the same DEM but using different algorithms. Figure 10.5b is the standard algorithm for aspect which uses Equation 10.8 to calculate slope in X and Y and Equation 10.6 to calculate aspect. Figure 10.5a simply takes aspect as the steepest downslope direction, and so is effectively making the calculation based on one neighbouring pixel instead of all eight. Figure 10.5c, which was derived using the Landserf package (www.landserf. org) calculates aspect from a polynomial surface fitted to the 5×5 pixels surrounding each point.

The algorithms differ both in the number of neighbouring points used and how they are used. All three capture the main changes in aspect on either side of the ridge and also the fact that the ridge top has been flattened by the interpolation process. However Figure 10.5a, being based on only one neighbouring pixel, is very sensitive to errors in the DEM and so the aspect values vary too much over a small area. Figure 10.5b is less variable and the pattern of aspect values is a better match with the slope shape as suggested by the contours. Figure 10.5c produces the smoothest result because it is not trying to fit the surface it uses exactly to all the points in the window. In effect it is

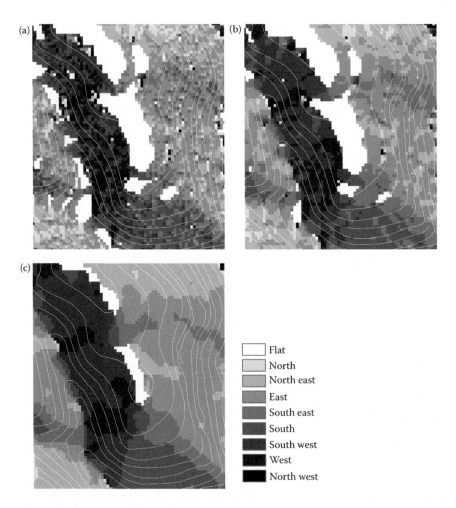

FIGURE 10.5
Slope aspect for part of the DEMs shown in Figure 9.20. (a) Steepest downslope direction. (b) Aspect using eight neighbouring pixels. (c) Aspect using 24 neighbouring pixels. (© Crown Copyright Ordnance Survey, all rights reserved.)

like the fifth-order polynomial in Figure 10.4 in that it is better at capturing the overall slope shape.

10.2 Hydrological Analysis Using a TIN

One of the major uses of surface models in GIS is the analysis of the flow of water across the landscape. Important tasks such as estimating the volume of water flowing in rivers and accumulating in reservoirs, assessing the risk of flooding and estimating soil erosion rates all require a knowledge of how water will flow across the terrain.

Central to this analysis is modelling what direction the water will take across the surface. Once this can be done for any point on the surface, all the other things we need to know can be based on this. For instance, when flow paths come together it can be assumed that these will become river courses. Given any point on the landscape we can trace the flow directions upslope to define the area which drains to that point – this area is called the catchment in the British hydrological literature, the watershed in the American literature.

As we know that water is driven by gravity, it will always follow the steepest path downslope, and so we can use this to model flow across our digital landscapes. Figure 10.6 shows part of a TIN representing a very simple river valley. It should be clear from an inspection of the heights of the triangle vertices that the triangles converge towards a series of edges which define the river valley that flows from right to left across the figure. This suggests one method of directing the flow across the TIN – assume that water will flow across each triangle to the lowest edge, and then flow along the edges, which effectively define the stream channels. This will work for triangles which directly border major streams, such as D and F in Figure 10.6. However, consider what this would mean for triangle C – water would flow to the lower edge, and then would all be forced to flow along this edge to the vertex marked 40, where it would meet the main valley.

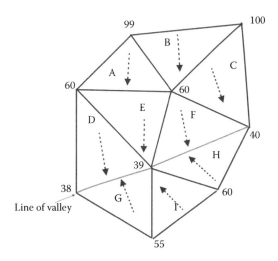

FIGURE 10.6
Flow across an imaginary TIN representing part of a river valley. The TIN edges representing the bottom of the valley are drawn in grey.

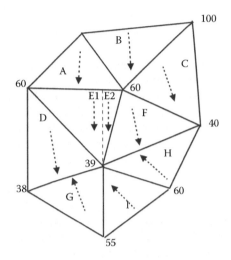

FIGURE 10.7
Subdividing flow across triangle E.

This simple approach would lead to some very strange-looking directions for drainage channels, and it is clear that to avoid this, we will have to allow water to flow from one triangle to another across edges. This means, for instance, that water from C will flow across F and hence to the valley. However, even this will produce some strange results. For instance, triangle E meets the stream at a vertex rather than an edge, and so flow from this triangle is forced to go into one of its neighbours (D or F) to reach the valley. This means that water falling in the upper right-hand corner of E is predicted to flow across both E and D, whereas water which falls nearby on the top of F will take a more direct route to the stream. One way to avoid this would be to subdivide the flow from triangle E into two parts as shown in Figure 10.7, with some being passed to D and some to F. This means that we will have to modify the original triangulation to include this new subdivision of triangle E.

These problems arise because of our assumption that channels can only exist along triangle edges. However, if we allow channels to direct water across the triangles, we run into further difficulties. In Figure 10.8, the flow

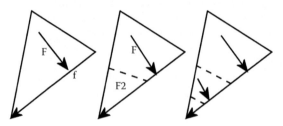

FIGURE 10.8
The problem of infinite regress in defining drainage channels on a TIN.

of water across F to the valley at the bottom is shown. If we view this flow as a single channel running down the triangle, then we have two channels meeting at the point f. In a real river network, channels are usually separated by a ridge, so two channels meeting on a plane surface like this is not very realistic. The implication is that there should be a ridge between them, as shown by the dotted line in the second diagram in Figure 10.8. However, this produces a new triangle (labelled F2). When we model flow across the surface of F2, we have the same problem as with F, and we will need to add a further ridge, producing another small triangle, and so on ad infinitum.

One solution to these problems is to use a triangulation which has been carefully constructed so that all the ridges and valleys in the landscape are properly modelled by triangle edges. In practice, many triangulations are produced by applying a standard Delaunay triangulation routine to a set of existing elevation data, and there can be no guarantee that the resulting triangulation will be a good model of the surface. Indeed, if the original data were from contours, there is good reason to believe that the triangulation will misrepresent key features of the landscape such as ridges and valleys.

10.3 Determining Flow Direction Using a Gridded DEM

The majority of GIS-based hydrological modelling uses the gridded DEM rather than the TIN. The main reason for this is the simplicity of handling the simple array data structure of a gridded DEM, compared with the complex TIN data structure. Some of the features of using the gridded DEM will be described in this section – for simplicity, the term DEM will be taken to mean a gridded DEM throughout.

The key to modelling the flow of water across the surface is to determine the direction in which water will flow from each DEM cell, which is done by considering the local 3×3 neighbourhood surrounding each grid point. The simplest method is to assume that water will flow to the neighbouring cell which has the steepest slope from the central cell. This will generally be the neighbour with the lowest elevation, but allowance has to be made for the fact that the diagonal neighbours are slightly more distant from the central cell than the others.

At first sight, this seems a reasonable approach since it represents the idea that water flows in the steepest downslope direction. However, it is not difficult to think of situations where the downslope grid point does not represent the main downslope direction from the centre point. Figure 10.9 shows a portion of a DEM that has an aspect of approximately 177° (i.e., very nearly due south). However, flow from the central cell will be to the southeast neighbour, a direction of 135°.

100	99	98
80	79	78
60	59	48

FIGURE 10.9
Imaginary DEM. The terrain is sloping almost due south, but flow from each cell will go to the south-eastern neighbour.

Strictly speaking of course, water will not flow from point to point across the surface, and in considering a local neighbourhood of grid points, such as in Figure 10.9, it is the flow of water from the cell surrounding each point which is being modelled. More sophisticated approaches are therefore based on two ideas:

1. That the flow from the cell all travels in the same direction, but that this is not constrained to be directly to one of the neighbouring cells. This means that unless flow *is* directly to a neighbouring cell, water may be distributed among several neighbouring cells.
2. That since the flow originates from an area rather than a point, it may not all travel in the same direction. In the case of flow from a hilltop, this might result in water being distributed to *all* of the neighbouring cells for instance.

The different flow direction algorithms will differ in the complexity of the calculations which need to be done for each grid point, but in terms of over-all complexity they will all be $O(n^2)$, as the calculations are done once for each grid point. A more telling comparison is that the single flow direction method is very sensitive to small errors in the DEM since a slight error in a single cell can completely alter the predicted direction of flow.

Once the direction from each cell has been estimated, the next step is to estimate for each cell how many other cells are 'upstream' of that point on the landscape. Cells which have a large catchment area can be assumed to be on the stream network, so a simple thresholding operation on the accumulated flow count will produce a predicted stream network.

The direction of flow from each cell is stored as a code number in that cell. As there are eight possible directions, it might seem sensible to store them as numbers from one to eight. In fact, in many systems, the numbers used are as shown in Figure 10.10.

The reason for this apparently strange set of numbers becomes clearer when they are each written out in binary form as shown in Figure 10.10. This shows that each number has a single 1 bit, with the rest of the bits being set

32	64	128
16		1
8	4	2

FIGURE 10.10
Codes used for directions out of a DEM cell.

to zero. There are two advantages to this. First, it is possible to store more than one flow direction from each cell. For instance, if water flows to both the south and the southwest, the code would be $4 + 8 = 12$ or 00001100 in binary. Because each of the eight codes only uses one bit, however many of them are added together, the new number will be unique.

The second advantage of this numbering scheme is that it allows a very efficient method of using the numbers to determine flow from cell to cell. The key to using the flow directions is to be able to tell, for any given cell, which of its eight neighbours contribute flow into it. For example, if the central cell in Figure 10.11 receives flow from the cell in the northwest corner, then the flow direction code in this northwest neighbour will be 2 – flow to the southeast. To determine how many of the neighbouring cells are sources of water, it is therefore necessary to look at each in turn and test whether it has the value shown in Figure 10.11.

The simplest way to do this is with a direct test for the number. In the pseudocode which follows, flowdir is an array which stores the flow directions in the eight neighbours of the cell being considered. Flowdir(nw) is the value in the northwest neighbour, flowdir(n) the value in the neighbour to the north, and so on. To count how many neighbours contribute flow to the current cell, the following set of tests would be used

```
1. if (flowdir(nw) == 2) n = n + 1
2. if (flowdir(n) == 4) n = n + 1
3. if (flowdir(ne) == 8) n = n + 1
```

etc.

2	4	8
1		16
128	64	32

FIGURE 10.11
Codes that represent flow into a cell from each neighbour.

Flow Direction Code Decimal	Flow Direction Code 8 Bit Binary
1	00000001
2	00000010
4	00000100
8	00001000
16	00010000
32	00100000
64	01000000
128	10000000

FIGURE 10.12
Flow directions codes in decimal and binary.

However, this method will become very cumbersome if we allow multiple flow directions from a cell. In this case, flow from the northwest cell could be represented not only by a value of 2 (flow to the southeast) but also by a value of 6 (flow to both south and southeast), for example. In fact, of the 256 possible flow direction values which could be stored in the northwest neighbour, half of them would include flow to the southeast. As the same is true of all eight neighbours, using the simple approach shown above would require 1024 separate if tests.

A more rapid method of performing this series of tests uses the fact that each direction is represented by a different bit in the flow direction code as shown in Figure 10.12. No matter what combination of directions are represented in the flow direction code, if there is flow to the southeast, the appropriate bit will be set to 1. The lower row of Figure 10.13 shows the number 2, which represents flow solely to the southeast, stored as an 8-bit binary integer. To help with the explanation, the individual bits have been numbered in the upper row of Figure 10.13. If there is flow to the southeast, then the number 1 bit will be set to 1 – if not, it will be a zero. If we can make this the leftmost bit in the byte, then there is a very rapid test we can use to see if it is a one or zero because in a signed byte, the leftmost bit is used to indicate whether a number is positive (leftmost bit 0) or negative (leftmost bit 1).

We have already seen in Chapter 8, that shifting bits within a byte is a very quick operation, so our test for whether a particular code includes flow to the southeast is simply:

1. if (leftshift(flowdir(nw),6) is negative) n = n + 1

Bit	7	6	5	4	3	2	1	0
Value	0	0	0	0	0	0	1	0

FIGURE 10.13
Storage of 2 in 8-bit byte.

Leftshift is a function which shifts all the bits in a byte left by the specified amount – 6 in this case. The same logic is applied to the other neighbours:

```
1. if (leftshift(flowdir(n),5) is negative) n = n + 1
2. if (leftshift(flowdir(ne),4) is negative) n = n + 1
```

etc.

This does not take any more time than the original code which tested for single flow directions – indeed, it may be a little quicker because left shifting and testing whether a number is positive or negative are both rapid operations.

10.4 Using the Flow Directions for Hydrological Analysis

The flow direction code is used in two of the most important surface operations – determining the size of the area upstream of a point and labelling the cells which lie in this watershed. Imagine that we wish to identify all the pixels which are upstream of the shaded pixel in Figure 10.14.

The method suggested by Marks et al. (1984) is an elegant one which makes use of the idea of recursion which was used for traversing the quadtree data structure in Chapter 8. In outline, the algorithm is very simple:

```
1. procedure wshed(pixel-id)
2. for each of 8 neighbours
3.    if IN watershed ignore
4.       if neighbour is upslope then
5.          mark pixel as IN
6.          call wshed(neighbour-id)
```

The algorithm starts with the watershed outflow point. In the pseudo-code, it is assumed that there is some way of labelling each pixel so that it is possible to keep track of which pixels have already been found to be in the

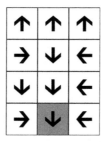

FIGURE 10.14
Flow directions in an imaginary DEM.

watershed. In practice, this would probably be done by using an array of the same size as the DEM itself, so that the pixel-id referred to would be the row and column number of the pixel in question. It visits each of the eight neighbours in turn. The first check is whether the pixel has already been dealt with, since pixels which are inside the watershed will on average be visited eight times. The next step is to check whether this neighbour flows into the original pixel, which is where the checking of the flow direction code comes in. If the neighbour does flow to the original pixel, it is marked as IN. At this point, the algorithm proceeds to check the eight neighbours of the pixel it is currently in by calling itself.

If the algorithm starts with the neighbour to the north, and proceeds clockwise, then after three iterations, the result will be as shown in Figure 10.15.

An outline of the sequence of operations might look like this:

```
1. Call wshed from start point
2. N neighbour IN watershed
3.    Call wshed from N neighbour
4.    N Neighbour IN watershed
5.       Call wshed from N Neighbour
6.       N neighbour not IN
7.       NE neighbour not IN
8.       E neighbour IN
9.          Call wshed from E neighbour
```

This algorithm is very efficient in processing terms because it only considers pixels which are in the watershed, or immediately adjacent. Each will be considered a constant number of times (eight) so the processing complexity is $O(k)$, where k is the number of pixels in the watershed. Note that the efficiency of this algorithm depends on the size of the output rather than the input, which is why the letter k is used instead of the letter n. If the watershed being identified has 50 pixels, it will take the same number of operations to identify these 50 pixels in a DEM of 100, 1000 or 100,000 pixels.

The algorithm is also very elegant and simple. The only drawback is that it might have large memory requirements. Every time wshed makes a call to

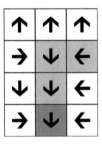

FIGURE 10.15
Watershed after three iterations of algorithm.

itself, the computer has to set aside extra memory to store the information which identifies which pixel is the 'current' one. For example, when called from the starting pixel, this might be identified as row 0, column 1. When the north neighbour is identified as belonging to the watershed, wshed is called again, this time with a new row and column number (1,1). These will have to be stored separately from the original 0,1 row and column numbers, because at some stage, the program has to come back and finish off going round the neighbours of this original point. This means that every time wshed calls itself recursively, two extra memory locations are needed. The maximum amount of extra memory required will depend on the size of the watershed, but in the worst case could be O(n^2), if the maximum dimension of the watershed is of the same order as the size of the raster layer.

Once the watershed has been determined, it is a very simple matter to determine its size, by counting the number of pixels which are IN, and this could be added to the above algorithm very easily. However, it is often necessary to determine what is called the flow accumulation, which is effectively the size of the watershed upstream from every pixel. If this were done by running the watershed algorithm for each pixel in the layer, this would have O($k * n^2$) complexity (remembering that we are taking n as the number of pixels along one edge of a raster layer, so that each layer has n^2 pixels).

The problem is that we cannot calculate the flow accumulation simply by looking at the immediate neighbours of each pixel since each of these may receive flow from an unknown number of other pixels. However, applying the watershed algorithm to each pixel is wasteful because neighbouring pixels will have watersheds which are very similar, so in effect we will be processing the same set of flow directions several times. An elegant solution to this problem suggested by Mark (1984) is to first sort the pixels into descending order of height. A pixel can only receive flow from pixels which are higher than it. If we process this new array in height order, it should be possible to deal with each pixel only once.

To see how this might operate in practice, we assume that we have already have a raster grid containing the pixel heights and a second grid containing the flow directions, as shown in Figure 10.16. The directions have been shown as arrows in the figure, for clarity, but in practice, the numerical codes described earlier would be used of course. The output of the algorithm will be a third grid in which each pixel contains the number of pixels which flow into it, which is also shown in Figure 10.16.

Let us assume for a moment that we have taken the heights from the DEM and stored them in a second array, which we have sorted into descending height order. The algorithm for assigning flow would then be as follows:

```
1.  Array HEIGHTS[1..N]
2.  Array FLOWS[1..N]
3.  Set FLOWS to 0
4.  for i = 1 to N
5.    Add FLOWS[i] + 1 to FLOWS[downslope neighbour]
```

(a)

147	127	124	137	167
131	110	108	122	153
114	94	92	107	140
98	79	77	93	126
82	63	62	79	114

(b)

↘	↓	↓	↙	↙
↘	↓	↓	↙	↙
↘	↓	↓	↙	↙
↘	↓	↓	↙	↙
↘	↓	↓	↙	←

(c)

0	0	0	0	0
0	2	2	1	0
0	4	5	1	0
0	6	8	1	0
0	8	11	1	0

FIGURE 10.16
Hydrological analysis. (a) Elevation values eights for a small part of a gridded DEM. (b) Flow directions. (c) Flow accumulation values.

Another array, called FLOWS, is set up to contain the flow accumulation value for each pixel. As each pixel is processed, it contributes 1 unit of flow to its downslope neighbour. It also contributes whatever flow it has received itself from its upslope neighbours.

Clearly, for the algorithm to work, we need some way of identifying the downslope neighbour of each pixel. This means there needs to be some way of linking the sorted array of heights back to the original DEM. The simplest is to store the row and column number along with the heights in a three-column array. The algorithm now becomes

```
1. Array DEM[0..maxrow,0..maxcol]
2. Array SORTED[0..nrows*ncols,3]
3. Array FLOWS[0..nrows*ncols]
4. for i = 1 to nrows
5.     for j = 1 to ncols
6.           SORTED[id,1] = DEM[i,j]
7.           SORTED[id,2] = i
8.           SORTED[id,3] = j
9. Sort SORTED array by first column
10.for i = 1 to nrows*ncols
11.    r = SORTED[i,2]
12.    c = SORTED[i,3]
13.    Add FLOWS[i] + 1 to FLOWS[downslope neighbour of DEM[r,c]]
```

Line 1 declares the size of the original DEM. In line 2, a second array is set up to store the heights and row and column ids. Note that since rows and columns are numbered from zero, the number of columns is maxcol + 1. Lines 4–8 then copy the heights and row and column IDs into SORTED. After the array has sorted, lines 10–13 process the array in height order. The key to this algorithm is line 9, in which the array of heights, row and column numbers is sorted. Sorting a multicolumn array like this is no problem, and is what is normally required of a sorting algorithm. In Chapter 6, a simple insertion sort was described which would run in $O(n \log n)$ time. The problem in this case is the potential size of the array to be sorted. It is not uncommon to deal with DEMs of 400×400 pixels, giving 160,000 height values. If we then add the same number of row and column ids, we have 480,000 numbers. Assuming each is stored in a 4-byte word, this gives a data size of just under 2 MB.

This is likely to present a problem for any sorting algorithm which needs to store the entire set of numbers to be sorted in memory. We know that the processing of information which is held in memory is much faster than the processing of information on secondary storage such as a disk drive. However, here, we have a situation in which it may simply not be possible to hold the entire array in memory. This is not a problem which is unique to spatial data of course. Commercial databases, such as customer account records, will often be too large to process entirely in memory and yet there will be a need to sort them, to facilitate searching for individual records. Therefore, alternative sorting algorithms have been developed which can deal with these large datasets.

One of the commonest is called the merge sort, and it is what is called a divide-and-conquer algorithm. We saw in the case of the insertion sort, and the Delaunay triangulation algorithm, that a useful way to design an algorithm was to try and simplify the problem. In those cases, the method was to deal with the data items one at a time. In divide-and-conquer algorithms, the problem is simplified by breaking it down into smaller and smaller problems. Eventually, the problem becomes so small that the solution is trivial. The second stage of the algorithm is to combine the solutions to all the simple subproblems to produce a solution to the original problem.

In the case of merge sort, imagine that we have 16 numbers to be sorted. Rather than sort all 16, we split them into two groups of 8 and try and sort each of these. If we repeat this subdivision, we will eventually have eight groups of two numbers. Sorting two numbers is a trivial operation. Now, we try and start combining our sorted groups of two numbers together. Imagine the numbers written on playing cards. On your left you have a pile of two cards, with the smallest on top and on your right a second pile, also with the smallest on top. To merge the two piles, you pick the smaller of the two top cards and place it on a third pile. You then repeat this process with the two cards which are now showing. If you exhaust one pile, then you simply transfer the rest of the other pile to the output. This procedure will work no matter how large the piles, as long as each is sorted.

Merge sort is ideal for large files because it is never necessary to process the entire file. In the first part of the algorithm, as much of the file as possible is read into memory, sorted and written out to a new file. The next part of the file is then processed in the same way to produce the second sorted file. When the sorted files are merged, they too can be processed a section at a time, since the records will be dealt with in order.

Merge sort is another algorithm which makes use of recursion in its design, as the following code of the algorithm, taken form Cormen et al. (1990) shows:

```
1. Procedure MERGE_SORT(A,p,r)
2. if p < r then
3.      q = p + r/2
4.      MERGE_SORT(A,p,q)
5.      MERGE_SORT(A,q + 1,r)
6.      MERGE(A,p,q,r)
```

A is an array containing elements numbered from p to r. The way this version of the algorithm is written, MERGE_SORT will split the range p to r into two halves and make a recursive call to itself until p is the same as r. When this happens, there will actually only be one element to be sorted, so MERGE_SORT exits without doing anything. When this has happened for both calls to MERGE_SORT, the MERGE procedure is called. This assumes that A[p..q] and A[q + 1..r] contain two sets of sorted numbers to be merged and returned in A[p..r].

To understand how this can sort anything, it may be helpful to see the sequence of calls which will be made if just four numbers are to be sorted. The first call will be to MERGE_SORT(A,1,4) – sort the first four elements of A which are numbered 1–4. The sequence will then proceed as follows:

```
1. MERGE_SORT(A,1,4)
2.      MERGE_SORT(A,1,2)
3.              MERGE_SORT(A,1,1)
4.              MERGE_SORT(A,2,2)
5.              MERGE(A,1,1,2)
6.      MERGE_SORT(A,3,4)
7.              MERGE_SORT(A,3,3)
8.              MERGE_SORT(A,4,4)
9.              MERGE(A,3,3,4)
10.     MERGE(A,1,2,4)
```

Remember that as long as there is more than one element to be sorted, MERGE_SORT will call itself twice and then call MERGE. The indentation in the pseudocode should make it clear that this happens for every call to MERGE_SORT except those on lines 3, 4, 7 and 8.

The MERGE procedure is a little more complicated since it has to process the two sets of numbers it is passed:

```
1. procedure MERGE(A,p,q,r)
2. Array MERGED[p..r]
3. left = p
4. right = q + 1
5. i = p
6. while left <= q and right <= r
7.      if A[left] < A[right] then
```

```
8.             MERGED[i] = A[left]
9.                left = left + 1
10.    else
11.             MERGED[i] = A[right]
12.                right = right + 1
13.    i = i + 1
14. if left > q then
15.    Copy A[right..r] to MERGED[i..r]
16. else
17.    Copy A[left..q] to MERGED[i..r]
18. Copy MERGED[p..r] to A[p..r]
```

This rather lengthy diversion makes it clear that it will be possible to sort the array of heights in our DEM flow accumulation algorithm. If p is the number of pixels in the DEM, the mergesort is an $O(p \log p)$ operation. Processing the sorted list is only $O(p)$ since each pixel is only dealt with once, and so the overall flow accumulation algorithm is $O(p \log p)$. (Note that this is a change from the previous cases where we have used the size of one side of a raster layer as the size of the problem, in which case the algorithm would be $O(n^2 \log n)$).

Further Reading

The book by Wilson and Gallant (2000) gives a comprehensive review of the calculation of many basic terrain parameters using both the gridded DEM and contour-based model. The standard reference for hydrological modelling techniques for gridded DEMs is Jenson and Domingue (1988) although both Hogg et al. (1993) and Tarboton (1997), in describing more sophisticated methods, also give a good overview of a range of algorithms in this area. Theobald and Goodchild (1990) describe some of the artefacts of TIN-based hydrological modelling, but conclude that these are the result of the way the TINs have been produced rather than any inherent weakness in using this data model for hydrological analysis. Van Kreveld (1997) discusses a range of surface algorithms using the TIN, including hydrological modelling, which is explored in more detail by Yu et al. (1996). Marc van Kreveld's homepage (http://www.staff.science.uu.nl/~kreve101/) contains links to online versions of these papers among others. Skidmore (1989) and Fisher (1993) were among the first to consider the possible effect that different algorithms might have on the results from the analysis of surfaces. Wise (1997, 2000b) has considered the effect of different interpolation algorithms in addition to different analytical algorithms on DEM results. More recently, similar studies have been done on LiDAR DEMs (Gallay et al. 2013, Leigh et al. 2009). Many of the traditional DEM operations as described in this chapter have been applied to LiDAR data but the 3D nature of the point data cloud which LiDAR generates means that many new applications are possible such as building detection and studying vegetation characteristics. For a recent review, see Liu (2008). Arge (1997) provides a very good discussion of the additional difficulties in designing algorithms for datasets which are too large to fit into memory.

11

Data Structures and Algorithms for Networks

As described in Chapter 9, although surfaces can be represented using both the vector and raster data models, most current GIS systems tend to use the raster model. This section will consider the example of networks, in which the reverse is true. Because the representation of networks is based on the fundamental data structures used for vector and raster data which have already been described, this chapter has a slightly different structure to the previous ones. The first section tries to indicate, in general terms, why the vector model is most often used for network applications. There are a number of important issues in the sorts of data structures which are used to represent networks, but these are most easily understood in the context of a specific algorithm, so the next part of the chapter focuses on methods for finding the shortest route between two points. This is a good example of an application which used to be quite specialist but that, owing to the growth in Internet mapping, GPS positioning and smart phones has become something which is used by large numbers of people on a daily basis. This means that algorithms have to be able to produce results quickly and some of the way in which this is achieved will be explored.

11.1 Networks in Vector and Raster

The networks that will be most familiar to many readers are road systems and river channels. Both are sets of lines, which as we saw in Chapter 1 (Section 1.2) can be stored in vector and raster. However, what distinguishes a network from other sets of lines, such as contour lines, is that the lines are connected and this has a crucial bearing on how they are represented in GIS. Figure 11.1 shows a section from an imaginary road network represented using both the vector and raster models.

The most obvious difference is the appearance of the two representations. The raster model gives a poorer representation of the road network because of the jagged appearance produced by the pixels. This can be improved, but only at the cost of a smaller pixel size, and hence larger files.

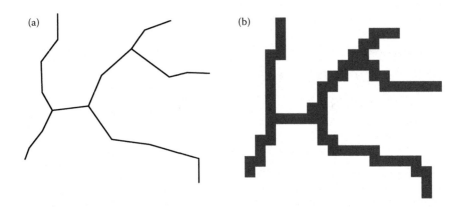

FIGURE 11.1
Imaginary road network represented in vector (a) and raster (b).

We also need to store attributes for each part of the network. These could be stored equally well in either model. With vector, each link would be stored as a line, approximated by a series of points. Each link would also have an ID, which would be used to associate the locational information with the attributes for each link stored in a database. The same approach could be used with raster. With many raster datasets, it is normal practice to store the attribute values directly in the pixels – this is how surface data would be stored, for example. However, in the case of networks, it is important to maintain the separate identity of each link as an object in its own right. If each pixel simply stores one of the attributes of the link, this separate identity is lost. However, there is no reason why the pixels cannot be used to store a link identifier instead, with the attributes stored in a related database. This would allow the system to answer a range of queries relating to the network such as identifying links which had particular attributes.

However, even with small pixels, and the ability to store the attributes of each link, the raster representation is less well suited to one of the main applications of road network data, which is the production of maps such as tourist route planning maps. For these maps, the link attributes are critical, as these are used to determine the symbolism used for the line. With a vector representation, the data provide the location of the course of the road, and vector graphics can be used to produce lines of different thicknesses, colours and so on. With a pixel representation of the route, it is more difficult to produce a range of different symbolisms. It is a very simple matter to change the colours of the pixels which gives a primitive level of cartographic symbolism. However, imagine trying to take part of the road shown in Figure 11.1, and represent it as a dashed line. It would not be possible to do this properly if each pixel is considered in isolation – to determine whether a pixel should be ON (i.e., drawn in black and hence part of a dash) or OFF (i.e., not drawn and hence part of the gap between dashes), it is necessary to know how far along the line each pixel is. This information is not stored in the raster data model, and would have to be calculated. In contrast, this sequence information is a fundamental part of the vector model, which stores each link as an ordered set of *XY* pairs.

The raster model is also less well suited to many of the analyses which are carried out using network information, such as finding routes. Planning routes through a network requires two types of data:

1. Information on the 'cost' of traveling along each link
2. Information on the connections between links

The simplest measure of how long it will take to travel along each road link will be its length. However, other factors will usually affect how long it takes to travel along a particular piece of road such as whether there are speed limits and what the volume of traffic is. These can all be generalised as a 'weight' or 'cost' attached to each road link. This information can be handled equally well in vector or raster, as it is just another attribute.

Information about the connections between links is somewhat more difficult to handle in raster than in vector. There are a number of ways in which connections between links can be modelled. For instance, a matrix can be drawn up with the links listed along both the rows and columns. Each cell in the matrix can store a 1 if the two links are connected, or a 0 if they are not. Alternatively, the cell might contain a number which represents the 'cost' of crossing that junction, as it can sometimes take additional time to cross the junction between two roads. However, a more flexible approach is to use the idea of a node, or a special point which is defined as the junction of two or more links. Nodes are a fundamental part of the link and node data structure used to represent lines in a vector GIS. However, in the raster model, the only spatial entity is the pixel. Even if a special code was used to identify certain pixels as nodes, the raster model has no natural way to represent which links this node connects.

None of these limitations of the raster model is insuperable. If the application needs to estimate travel costs or least-cost routes across a whole surface, and not just along a network, then the raster model has considerable advantages over the vector model. However, for applications which need to model flow along a network, the vector model is more straightforward to use. The remainder of the discussion of algorithms and data structures in this chapter will therefore concentrate on the vector model of networks.

11.2 Shortest Path Algorithm

One of the most fundamental tasks in network analysis is to find the shortest path between two points. To illustrate the basic algorithm used to solve

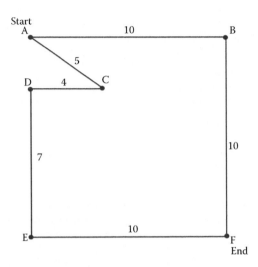

FIGURE 11.2
Simple network.

this problem, we will use the simple network shown in Figure 11.2 – the task is to find the shortest route from point A to point F. Note that the figures against each link are simply the distance, in arbitrary units, between the nodes. However, they can represent anything we wish about the link and can be more generally thought of as a weight associated with the link. In a road network, the weight might be related to the time taken to travel along the link, which means that the weight from A to B might be different from that from B to A. If A was at the top of a hill and B at the bottom, the AB time might be less than the BA time for instance.

This is rather similar to the very first problem which was posed in the introduction – finding a route through a maze. It is very simple for us to see at a glance that there are only two routes from A to F – one via B, which has a distance of 20, and one via C, D and E, which has a distance of 26. However, the computer of course cannot 'see' the whole network – it has all information about the nodes and the links between them and will need an algorithm to solve the problem. In fact, in a real-world application of this problem, such as planning routes for going on holiday or transporting goods, seeing the whole network would not necessarily allow us to arrive at the best solution.

As with most problems, there is a brute force approach, and as with most brute force approaches, it will work, but will not be very efficient. In this case, the brute force method is to find all possible routes from A to F, calculate the total distance of each, and select the shortest. With the network as it is, this would be quite simple, since there are only two possible routes:

Route 1: A, B, F

Route 2: A, C, D, E, F

If we add the link between C and F, this adds one more route:

Route 3: A, C, F

However, if we add a link between B and C as well, this adds several more routes because we now have a greater choice of routes out of C. The new A–F routes are

Route 4: A, B, C, F – a modification to route 1 to go via C

Route 5: A, B, C, D, E, F – a modification of route 2 to go via B (Figure 11.3).

As the network size grows, each new link produces a new variation on some proportion of the existing links, so the bigger the network, the more potential routes each extra link produces. This type of growth, in which the rate of growth is proportional to the current size of the problem, is called exponential and Figure 11.4 provides some indication of how quickly exponential problems can grow. The column labelled quadratic shows how the problem size grows in the case of $O(n^2)$ complexity, which we have already come across in the context of other algorithms. In the case of exponential complexity, the exponent in the equation is not fixed at 2, as with quadratic growth, but is itself a function of the problem size n – by the time n has reached 7, the exponential problem is already an order of magnitude larger than the quadratic one.

Trying all possible routes is clearly not an option here. The classic algorithm is called the Dijkstra (pronounced Dike-struh) algorithm, after the computer scientist who first described it. It is based on finding the node which when added to the current route will create the shortest path back to

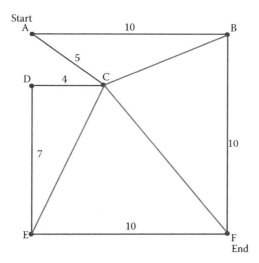

FIGURE 11.3
Network from Figure 11.2 with extra links added.

n	Quadratic n^2	Exponential x^n
2	4	4
3	9	8
4	16	16
5	25	32
6	36	64
7	49	128
8	64	256
9	81	512
10	100	1024
11	121	2048
12	144	4096
13	169	8192

FIGURE 11.4
Comparison of quadratic and exponential growth.

the start point. It can be used to find the shortest route from every node in the network back to the start or it can stop when it has found the route to a particular end point. If applied to Figure 11.2, it would start at node A and look at the distance to all the neighbouring nodes. The shortest is the link to C so this is added to the route. From C, D will be picked giving a distance of 9 units from the origin. However, the link to E adds a further 7 units, which gives a distance of 16 from the origin. This is further than the direct route from A to B, and so this would be selected in preference. Because the shortest route back to the origin is selected at each stage, when the destination is reached, the route found must by definition be the shortest. In fact, this can be proved more rigorously than this, and further reading is given at the end of the chapter for anyone who wants to follow this up.

It is worth describing in more detail how this algorithm is implemented in practice because it illustrates some important points about how the network is stored as a data structure. This description is based on that given by Worboys and Duckham (2004) – the reading list provides details of other alternative accounts and even of some animations of the algorithm on the World Wide Web. For the moment, let us assume that for each node, we can work out which links start at that node and how long they are. We set up three arrays of information for each node:

- *Distance*: The total distance from this node back to the origin. This is set to zero for the start node and infinity for all other nodes.

- *Parent*: The next node back along the route to the origin. This is set to blank for all nodes.

- *Included*: Whether or not this node has ever been included as part of the shortest route. This is set to NO for all nodes.

Node	Distance	Parent	Included
A	0	–	NO
B	∞	–	NO
C	∞	–	NO
D	∞	–	NO
E	∞	–	NO
F	∞	–	NO

FIGURE 11.5
Table at the start of the algorithm.

At the start of the algorithm, the arrays are as shown in Figure 11.5. The algorithm is described below. The notation dist(N) indicates the current value of the distance column for node N. The notation d(NM) indicates the distance between nodes N and M.

- Look at all nodes which have not been included so far (whether or not they have a link with the current position) and choose the one with the shortest current distance back to the start.
- Mark it as YES in the included array. Let us call it N. The distance from this back to the start is dist(N).
- Find the nodes which are connected to N and which are still marked NO in included. For each of these M nodes, the distance between it and N is d(NM).

```
1. for each node
2.    if dist(N) + d(NM) < dist(M) then
3.       dist(M) = dist(N) + d(NM)
4.       parent(M) = N
```

In other words, if there is a shorter route back to the start from node M via node N, then the arrays are updated to record this fact.

In the first iteration, the node with the smallest distance value is A, since it has a distance of 0. This is picked and its included value set to YES. Two nodes are connected to A – B and C. Their current distances back to the start are infinity. In both cases the route back to A is shorter than this, so their distance values are updated, and their parent node is set to A. This produces the arrays shown in Figure 11.6.

In the second iteration, nodes B through F are considered and C found to have the shortest current distance back to the start. The included column is now updated to YES for C, so it will no longer be considered in the algorithm. It only has one neighbour, D, which is 4 units away from C. Adding this distance, to the distance from C back to the start (5) gives a distance from D back to the start of 9 units. This is less than the current distance for

Node	Distance	Parent	Included
A	0	–	YES
B	10	A	NO
C	5	A	NO
D	∞	–	NO
E	∞	–	NO
F	∞	–	NO

FIGURE 11.6
Table after one iteration.

D, which is still infinity, so the distance column for D is updated with a value of 9 and its parent is set to C (Figure 11.7).

In the next iteration, D is the node with the shortest current route back to the start. It only has one neighbour, E, and so after this iteration, the table will be as shown in Figure 11.8.

From E, there is a direct route to F, but we know that this would not be the shortest. What the algorithm does is to pick node B next, since this has not so far been included on the route, but now has the shortest route back to

Node	Distance	Parent	Included
A	0	–	YES
B	10	A	NO
C	5	A	YES
D	9	C	NO
E	∞	–	NO
F	∞	–	NO

FIGURE 11.7
Table after two iterations.

Node	Distance	Parent	Included
A	0	–	YES
B	10	A	NO
C	5	A	YES
D	9	C	YES
E	16	D	NO
F	∞	–	NO

FIGURE 11.8
Table after three iterations.

Node	Distance	Parent	Included
A	0	–	YES
B	10	A	YES
C	5	A	YES
D	9	C	YES
E	16	D	NO
F	20	B	NO

FIGURE 11.9
Table after four iterations.

the start. B only has one neighbour, which has not yet been included, F. The distance BF plus the distance back to the start from B are added together and put into the distance column for F (Figure 11.9).

The next node to be considered will be E since of the two nodes which have not had their included value set to YES, this has the shorter distance back to the origin. The distance from E to the start is 16. The distance from E to its only non-included neighbour is 10. Since the sum of these two is greater than the current distance back from F to the start, no change is made to the distance column for F. After the fifth iteration, the table looks as shown in Figure 11.10.

On the next iteration, the only node left to be considered is the destination, and this is where the algorithm terminates. The final table shows that the shortest path from F back to A is 20 units long. The actual route can be determined by tracing back through the parents – from F back to B and hence to A. Another interesting feature of the Dijkstra algorithm is that the final table actually contains the shortest distance and route from every node back to the start point.

The efficiency of the algorithm depends on the number of nodes (n) and the number of links (l) in the network. On each iteration of the algorithm, one node is added to the list of those INCLUDED. As this happens only once for each node, there are O(n) iterations of this main loop. Each link has a

Node	Distance	Parent	Included
A	0	–	YES
B	10	A	YES
C	5	A	YES
D	9	C	YES
E	16	D	YES
F	20	B	NO

FIGURE 11.10
Table after five iterations.

distance associated with it. This is considered just once – when one of the nodes which it joins is added to INCLUDED. When the node at the other end is later added to INCLUDED, the distance is not considered again. Hence, each link is considered once making the total complexity $O(n + 1)$. However, there is one further crucial element to consider. On each of the n iterations of the main loop, the node with the shortest distance back to the origin, which has not already been considered, has to be selected. If the nodes are stored in a simple array, as shown in the tables above, the only way of finding the one with the shortest distance is to search the whole array, which is an $O(n)$ operation. Since this is done on each of the n loops, this makes the total complexity of the algorithm $O(n^2 + 1)$, which is $O(n^2)$.

So far we have simply assumed that we can derive all the information we need to know about the network to operate the algorithm. In the next section, we will consider what data structures we will actually need to do this, and from this we will see how we can improve the efficiency of this algorithm from $O(n^2)$ to $O(n \log n)$.

11.3 Data Structures for Network Data

We have already covered the storage of lines and the connections between them in describing the basic vector data structures. Figure 11.11 shows the same basic network, but with each link represented by a number, rather than its weight.

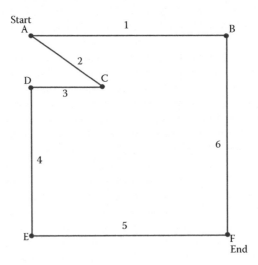

FIGURE 11.11
Simple network with link identifiers.

Figure 11.12 shows how this network could be represented in a table based on the link and node data structure which was introduced in Chapter 3. The question is whether we can use this table to fill in the three arrays we need for the Dijkstra algorithm. The algorithm consists of three steps, which can be simplified to

1. From the nodes which have not been included, pick the one with the shortest value in the distance array.
2. Mark this as included and note its current value in the distance array.
3. Find the nodes which are connected to this one. For each connection, find the weight assigned to the link. See if this provides a shorter route back to the start than the current one.

The first two steps can be handled entirely from the distance and included arrays, although we will see later in this chapter that this is not the most efficient way to handle these steps. First, let us consider the third step and start by considering the situation when node C has been picked as the current node. We now need to find out

1. Which nodes have a direct link to C?
2. Which of them have already been included?
3. Of those which have not, what is the weight of the link from node C?

All this information can be supplied using the table in Figure 11.12. To answer the first query, we would need to search the table and extract all rows which had C as either the start or end node. In the context of the current operation, we do not care whether a link goes from A to C or from C to A. This would pull out the records for links 2 and 3. Now we note which node is at the other end from C in each case. We check in the included table to see which have already been considered. This will leave only link 3, as node

Link ID	From node	To node	Weight
1	A	B	10
2	A	C	5
3	C	D	4
4	D	E	7
5	E	F	10
6	B	F	10

FIGURE 11.12
Link and node representation of simple network.

A has already been dealt with. From the record for link 3, we can read the weight which provides the answer to the third query.

This seems fine in this simple case. However, consider what would happen if instead of six links we were dealing with 600, 6000 or even 60,000. The first of our three queries above now becomes a matter of selecting two links from a total of 60,000. If we simply start at the first row and search a row at a time, this is going to be very inefficient, especially when we consider that this is only part of an algorithm which will be repeating this operation at least once for every node in the network. It is clearly worth trying to find a quicker way to answer this query.

One option is to create a second table, which lists the links that are connected to each node (Figure 11.13).

This table can be created in $O(n)$ time from the table in Figure 11.12. Now, our query becomes much quicker because we can go directly to the correct entry in the node table, as we know which node we are looking for. The entry will tell us which links we need to look at, so we can extract these directly from the link table – there is no need to search every link record.

In fact, the basic table we used for the Dijkstra algorithm also had a record for each node, so rather than have an extra table it makes sense to add the information about the links connected to each node to this table. The problem with the table in Figure 11.13 is that different nodes may have different numbers of links connected to them, which is messy to deal with. One solution to this is to store the information in two linked tables as shown in Figure 11.14.

The node table has been modified to provide a pointer to one of the links which starts at that node, and one of the ones which ends there. In the link table, if more than one links starts at the same node, the record for the first link contains a pointer to the next link. If there are no further links, then the pointer contains –1. To find out which nodes neighbour C, we can use these two tables to search through two lists in the link table – the list of links which start at C and the list of those which terminate at C.

Node	Links
A	1,2
B	1,6
C	2,3
D	3,4
E	4,5
F	5,6

FIGURE 11.13
Node table for network.

LINK Table

Link ID	From node	To node	Weight	Next link from start	Next link to end
1	A	B	10	2	−1
2	A	C	5	−1	−1
3	C	D	4	−1	−1
4	D	E	7	−1	−1
5	E	F	10	−1	6
6	B	F	10	−1	−1

NODE Table

Node	Distance	Parent	Included	First link out	First link in
A	0	—	NO	1	−1
B	∞	—	NO	6	1
C	∞	—	NO	3	2
D	∞	—	NO	4	3
E	∞	—	NO	5	4
F	∞	—	NO	−1	5

FIGURE 11.14
Storage of network information using separate node and link tables.

This data structure will speed up the operation of the Dijkstra algorithm by making the search for neighbouring nodes faster than if we searched the original table of information for the links. However, it will not improve the overall efficiency of the algorithm, which is still $O(n^2)$ because we process each node once, and in each case we need to find the node with the shortest distance back to the origin which has not so far been considered. If we store the distance in a simple array, the second step will be an $O(n)$ operation. However, if we can sort the distances into order, we will be able to find the shortest distance more quickly. In fact, if the distances are sorted into order, the shortest is the first in the list so retrieving it becomes $O(1)$. However, this is not simply a matter of taking an array and sorting it, which we know can be done in $O(n \log n)$ time. The values in the array will change, as the distances are updated, and the elements will be removed from consideration when they are added to the INCLUDED list. This means we need some way of updating our sorted list as changes are made, without having to completely resort it. A data structure which can do this is called a priority queue, and a common form of priority queue is the binary heap.

The binary heap is a tree data structure, like the quadtree which was described in Chapters 7 and 8, but one in which each node can only have up to two children, rather than four. The binary heap has three properties:

- The children are both equal to or larger than their parent.
- Each node in the tree has a maximum of two children (hence the binary in the name).
- The tree is constructed and maintained so that each successive layer of leaf nodes is filled before the next is begun.

The first property means that the root node of the heap contains the smallest value, which is why it is useful as a priority queue. Like all trees, many operations using the tree can be done O(log n) time, but the second and third properties make the binary heap particularly efficient in terms of both storage and time.

To illustrate how the binary heap works, let us start with the example shown in Figure 11.15 which shows a heap containing 10 entries. The values shown inside the circles are the distances back to the origin in the Dijkstra algorithm. For the moment, we will assume that we know which network location each of these distances relates to, without worrying how this might be done.

The tree is what is called a full tree because when it is built, each level is completely filled by giving each node two children, apart from the lowest level, which is filled from left to right as shown in Figure 11.15. If you check, you will see that each node is smaller than its two child nodes. Note, however, that this heap property only applies to nodes and their immediate children. For instance, node 2 with a value of 8 is actually larger than node 6, which is in the next layer down and has a value of 7. However, because they are not directly connected in the tree, this does not matter.

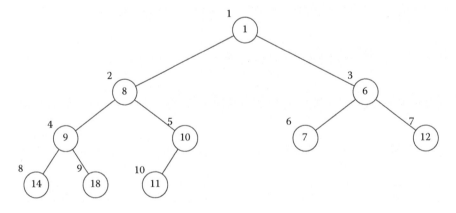

FIGURE 11.15

Heap. The figures outside the nodes are the node identifiers (1–10). The figures inside the circles are the numbers held in each node.

To see how the tree is used, let us assume the location referred to in node 1 is chosen as the next location on the shortest path. This means that this entry is removed from the tree and we need to replace it with the smallest value among the nodes which are left. One way to do this would be to select the smaller of the children of node 1, which would be node 3. The 6 would be moved from node 3 to node 1, and then the 7 moved from 6 to 3. This would give the tree shown in Figure 11.16.

However, this is no longer a full tree because the second layer is no longer full. Having a full tree makes it much easier to store the tree in an array because the number of nodes in each layer of the tree is predictable – 1, 2, 4, 8 and so on. Therefore, as long as we store the node values in sequence in the array, we know exactly where the value for any given node is stored. It is therefore important that we have a method of replacing the value in the root, which ensures that the tree remains full.

Once the root node is empty the tree will only contain 9 values and in order to maintain the property of the tree being full, the empty node will need to be the last one – node 10. Therefore, we move the value from node 10 to the root node which gives us the tree shown in Figure 11.17.

We now have a tree which violates the heap property because node 1 is larger than both its children. So, the next stage is to restore the heap property by using a procedure called heapify, which looks like this:

```
heapify (startnode)
Find which is smallest from startnode, left-child, right-
child.
If startnode not smallest
    Swap smallest with startnode
    heapify(smallest)
```

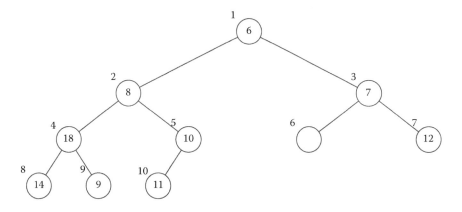

FIGURE 11.16
Heap after moving smallest child into vacant root node and restoring the heap property to the tree.

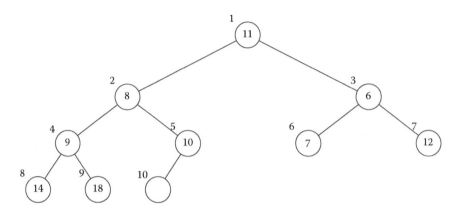

FIGURE 11.17
Heap after moving 11 from node 10 to the root node.

This will start at node 1 and examine the values in nodes 1, 2 and 3. As node 1, which is the startnode, is not the smallest, it will be swapped with the smallest, giving the tree shown in Figure 11.18.

We now have the smallest value in the root of the tree, but the newly swapped value in node 3 is larger than one of its child nodes. Because there is always the chance that this will happen, heapify always calls itself again starting with the node which has just been swapped. When heapify is called again, the values in nodes 3 and 6 will be swapped and we will get the tree shown in Figure 11.19.

The tree has now been properly restored but the current version of heapify has not been properly coded to allow for this, so let us look at a fuller implementation of the procedure. The first thing to consider is how we identify the left and right child of any node.

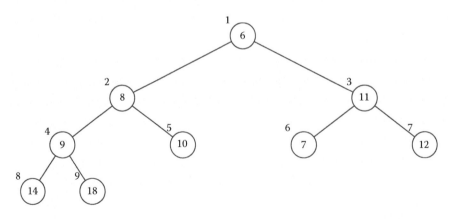

FIGURE 11.18
Heap after swap of root node with its smallest child.

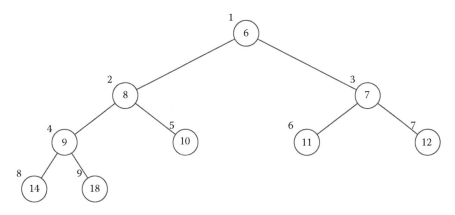

FIGURE 11.19
Heap with heap property restored.

As said earlier, because the heap is always a full tree, it can be stored in a simple array as shown in Figure 11.20. The root will always be stored in element 1 of the array. The two nodes in the next level of the tree will be in entries 2 and 3, and the four in the next level will be in 4–7.

Not only are these positions fixed but also there is a very simple way to find the location of the two child nodes for any node in the tree. If the node is entry i in the array, then

```
left(i) = i*2
right(i) = i*2+1
```

You can check for yourself by looking at the node numbers in any of the diagrams, but remember – it is the node IDs which are being referred to, not the values held in the nodes. Finding the parent of node is equally easy:

```
parent(i) = i/2
```

Note that this uses integer division, in which the remainder is ignored. This means that $5/2 = 2$ and not 2.5. We also have the benefit that multiplication and division by 2 are very efficient on computers, as they can be accomplished simply by shifting the bits in a number one position to the left or right, respectively.

Array index	1	2	3	4	5	6	7	8	9
Value	11	8	6	9	10	7	12	14	18

FIGURE 11.20
Storage of the binary heap from Figure 11.19 in an array.

With this understanding of how the heap is stored and accessed, we can present a more detailed version of Heapify, taken from Cormen et al. (1990):

```
heapify (A,i)
/* A is the array which contains the heap
/* i is the index value of the first node to be considered
/* Find ID of child nodes
l = left(i)
r = right(i)
/* Find ID of which has the smallest value
if l <= heap_size and A[l] < A[i]
    then smallest = l
    else smallest = I
if r <= heap_size and A[r] < A[smallest]
    smallest = r
/* See if current node is smallest
/* If not swap its value with the smallest
/* and call HEAPIFY starting at swapped child
if i != smallest
    temp = A[i]
    A[i] = A[smallest]
    A[smallest] = temp
    heapify (A, smallest)
Procedure left(i)
Return i*2
Procedure right(i)
Return i*2 + 1
```

If heapify is run starting from the root node, the maximum number of iterations will be determined by the number of levels in the tree. Because the tree is binary, the maximum number of levels is log n and heapify will run in O(log n) time. This means that the Dijkstra algorithm implemented using a binary heap as a priority queue will also run in O(n log n) time.

There is a slight complication in the use of a binary heap. The examples so far have only shown distances stored in the heap. However, we would need to store the actual ID of the node in the network along with its current distance value. This could be done by using two positions in the heap array for every node in the heap, and altering the left() and right() functions accordingly. However, we also need to be able to find the position of any given network node in the heap, so that we can update its distance value. The heap is not an efficient structure for this type of query and this would necessitate a linear search through the heap array. To avoid this, we would add an additional pointer to our node table, which pointed to the position of each node's record in the heap array. This would have to be updated every time we modified the heap of course, but this is an O(1) operation, so it has no effect on the overall complexity of our algorithm.

This section has further developed the idea that the design of data structures is just as important as the design of algorithms in the development of efficient computer programs. We will end our consideration of networks with an example of a problem for which no efficient algorithms are known to exist, and where alternative strategies have to be adopted.

11.4 Faster Algorithms for Finding the Shortest Route

Let us now return to the original problem of how to find the shortest route between two points. The Dijkstra algorithm will solve this problem and it can be proved that it will always find the shortest route. The problem is that because of the way it works it spends a lot of time processing parts of the road network which cannot possibly lie on the desired route. Because it is based on selecting nodes which produce the shortest route back to the start, this means that if the final route is 50 km long, it will have to consider all nodes which lie within 50 km of the start. To illustrate what this means, Figure 11.21 shows the approximate area which would be searched by the algorithm in finding a route between Sheffield and Manchester, which are 52.2 km apart.

As the figure illustrates very graphically, the algorithm could be speeded up if the search could be restricted to roads leading in roughly the right

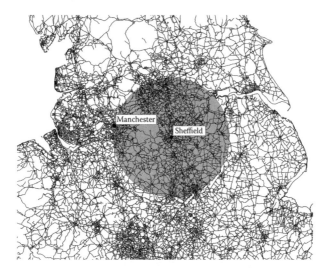

FIGURE 11.21
Roads which would be considered by the Dijkstra algorithm in finding the shortest route between Sheffield and Manchester in the North of England. (Crown Copyright Ordnance Survey, all rights reserved.)

direction. When route planning was something which was done in the offices of haulage companies when planning journeys for their fleet, the speed of the algorithm was not so crucial. However, route planning is now something which people routinely do during journeys using satellite navigation software in their cars or on their phones and so the algorithms need to be fast.

One of the commonest algorithms is called A* (pronounced A-star). The A* algorithm differs from the Dijkstra algorithm in two ways:

1. At each step, instead of searching all the remaining nodes to see which should be used next, it only searches those with a direct connection to the current node.
2. In deciding which node to add to the current route, the algorithm takes into account how near each one is to the destination as well as how far it is back to the origin.

The first change alone will make a large difference to the speed of the algorithm as most nodes in a transport network have direct links to only three or four other nodes. It is rare to find a road junction at which more than four roads meet. This means that at each step of the algorithm, instead of searching what could be hundreds of nodes, the algorithm only considers a few. To understand how the second condition is implemented, it is useful to introduce the notation which is usually used to describe the various route-finding algorithms. When choosing the next node, the Dijkstra algorithm calculates a function, $g(n)$, for each candidate node which is made up of the distance to the current node along the route, and the distance from the current node to node n. In notation

$$g(n) = g(\text{current}) + \text{dist}(\text{current}, n)$$

where $g(\text{current})$ is the distance from the start to the current node, $\text{dist}(\text{current}, n)$ is the distance between the current node and the node n which is being evaluated.

The other methods that we will consider all evaluate this $g(n)$ function and also other factors, so, in general, we can think of all these methods evaluating a function $f(n)$, where for the Dijkstra algorithm

$$f(n) = g(n)$$

and for A*

$$f(n) = g(n) + h(n)$$

where $h(n)$ is an estimate of how far it is from node n to the destination.

As we will see, this is a useful notation because it distinguishes things which can be measured – $g(n)$ – from things which must be estimated – $h(n)$. It is clearly a sensible strategy to choose a node which moves us closer to our destination, but the problem is that until we have found the route, we do not know how far each node is from the destination along the best route. However, we do know how far each node is from the destination in a straight line, and so we can use this as an estimate for $h(n)$. Such an estimate is called a heuristic estimate and can be thought of as an educated guess at the correct answer. The $h(n)$ values for each node are shown in Figure 11.22.

Straight line distance has two advantages as a heuristic:

- It is very easy to calculate as we know the location of both points.
- In practice, it is actually quite a good estimator of the road distance (Phibbs and Luft 1995).

Let us see how using the heuristic values modifies the selection of the route. First, some new terminology will be introduced. So far, it has been assumed that we are trying to find a route along a real road network. However, this is simply one example of a more general problem of finding the least-cost route through a graph. Graph theory was introduced in Chapter 3 as one of the roots of vector data structures. However, graphs, as well as modelling maps, can be used to represent all sorts of problems. Some of these involve real networks such as transport networks. However, in some, the graph is being used to represent other sorts of connections, such as the links between different stages of a project in critical path analysis. In the more general terminology of graph theory, the starting node is usually referred to as the origin

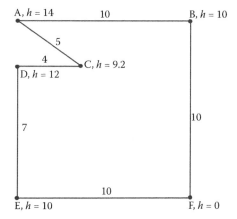

FIGURE 11.22
Network with *h* values shown. In this, the heuristic is the straight line distance to node f.

and the destination as the goal. At any point in the algorithm, there are two sets of nodes:

- The open set – those nodes which are under active consideration as candidates for being part of the solution.
- The closed set – those nodes which have been considered and will not be looked at again. Note that some of these will be on the final route and some will not.

First we will take a look at the Dijkstra algorithm again using this new terminology and then compare this with the A-star algorithm. The Dijkstra algorithm starts with the open and closed sets populated as follows:

```
Open set = {All nodes except for the origin}
Closed set = {The origin}
```

Nodes are moved to the closed set once their distance value has been updated from its initial value of infinity and labelled with their parent node. The full algorithm as applied to the example network is shown in Figure 11.23.

The A-star algorithm begins with the two sets as follows:

```
Open set = {Origin}
Closed set = {The empty set}
```

Step 1	Open set: {A,0; B,INF; C,INF; D,INF; E,INF; F,INF} Closed set {} Move A to closed set. Update B and C.
Step 2	Open set: {B,10; C,5; D,INF; E,INF; F,INF} Closed set {A} Move C to closed set. Update D.
Step 3	Open set: {B,10; D,9; E,INF; F,INF} Closed set {A,C} Move D to closed set. Update E.
Step 4	Open set: {B,10; E,16; F,INF} Closed set {A,C,D} Move B to closed set. Update F.
Step 5	Open set: {E,16; F,20} Closed set {A,C,D,B} Move E to closed set. F via E = 26. Leave F.
Step 6	Open set: {F,20} Closed set {A,C,D,B,E} Move F to closed set. Algorithm terminates.

FIGURE 11.23

The steps in the Dijkstra algorithm for finding the least-cost route from node A to node F in Figure 11.22. The nodes in the open set are shown with their distance values.

The algorithm works as follows:

1. The node in the open set with the lowest *f(n)* is selected and moved to the closed set along with a note of its parent node.
2. The neighbours of the current node have their *f(n)* values calculated and are moved to the open set.

These two steps are repeated until the node which is selected at step 1 is the goal, at which point the algorithm terminates. Let us see how the algorithm would work for the network in Figure 11.22.

The algorithm begins at the start node A and considers its two neighbours – B and C. Their scores are calculated as follows:

```
g(node) = g(A) + distance(A, node)
f(node) = g(node) + h(node)
```

In the case of B and C, these calculations give the following:

```
g(B) = 0 + 10 = 10
g(C) = 0 + 5 = 5

f(B) = 10 + 10 = 20
f(C) = 5 + 9.2 = 14.2
```

Node A is added to the closed set and B and C are added to the open set.

On the second time through the nodes in the open set are searched and the one with the smallest *f* score is selected. This is node C and so this is moved from the open to the closed set. The neighbours of C are then examined to calculate or update their *f* and *g* scores. Node A is in the closed set and so is not considered. Node D has not been considered so far and so its *f* score is calculated:

```
g(D) = g(C) + distance(C,D) = 5 + 4 = 9
f(D) = g(D) + h(D) = 9 + 12 = 21
```

At this stage of the algorithm, the open and closed sets look as follows:

```
Closed set = {A,NULL;C,A}
Open set = {B,20;D,21}
```

The nodes in the closed set have their parent node listed. This is NULL in the case of A as this is the origin.

At the next step, node B would be selected for consideration since it has the lowest *f* score of the nodes in the open set. B would be added to the closed set and its neighbours considered. This means that the scores for the destination node F would be calculated:

```
g(F) = g(B) + distance(B,F) = 10 + 10 = 20
f(F) = g(F) + h(F) = 20 + 0 = 20
```

The sets now look like this:

```
Closed set = {A,NULL;C,A;B,A}
Open set = {D,21;F,20}
```

At the next step, F would become the current node and the algorithm would terminate.

In comparison with the Dijkstra algorithm, A* has taken four steps instead of seven. This may not seem like a large difference, but that is because the network is so simple that there are not that many possible routes to consider. To see why A* will be faster for real networks, consider the network shown in Figure 11.24, in which the original network has been expanded by adding in some additional roads around the start point.

These will have a minimal effect on the A* algorithm. The neighbours of A and B will be added to the open set, but will never be selected for further consideration because they are further away from the destination than any of the other nodes. However, many of the new nodes are closer to the start than node F and so will be considered by the Dijkstra algorithm.

As well as being fast, the A* algorithm possesses some other useful characteristics. If the distance to the goal is used as the heuristic, then the algorithm is guaranteed to find the least-cost route to the goal. In addition, once

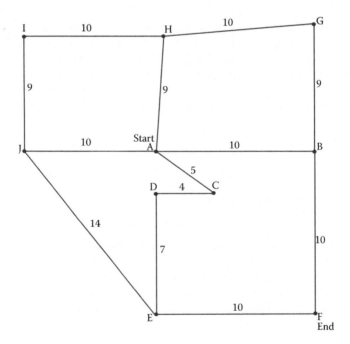

FIGURE 11.24
Expanded road network.

a node has been selected from the open set, then the lowest-cost route to it has already been found and it does not need to be considered again. This is why the algorithm can stop as soon as the destination is selected as the current node. But why is this dependent on using distance as the heuristic? To understand this, it is easiest to use a very simple example of part of a network, which consists of just three nodes as shown in Figure 11.25.

In Figure 11.25a, distance is used as the heuristic. For the A* algorithm to produce an optimal route, the heuristic must have two properties:

1. It must be what is called 'admissible'. This means that it must not overestimate the actual distance from a node to the goal.

2. It must obey the triangle inequality, which says that the length of any side of a triangle must be less than the sum of the lengths of the other two. This means that the distance from node A direct to node G in Figure 11.25a must be less than the distance from A to G via node B.

As long as distance is used for the heuristic, these two conditions are satisfied. The straight line between two points is the shortest distance between them. In fact, this is one of the fundamental axioms of geometry and dates all the way back to Euclid. So whatever the true distance along the road between A and G, it cannot be less than 10. The second rule also holds true for straight line distances because these can always be represented as a triangle on a plane.

But what if we do not use distance as the heuristic? What if we use travel time instead? In planning routes, we know that different roads have different average speeds and so most route planners offer the option of finding the fastest route rather than the shortest. There is no problem with doing this because all we need to do is use estimates of travel time as the heuristic as shown in Figure 11.25b. The problem is that our two rules no longer apply. The time to get from A to G is now longer than the time going via B. And there is no guarantee that our estimate of 10 as the time to get from A to G

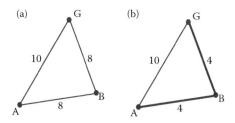

FIGURE 11.25
Different heuristics which can be used for the A-star algorithm. (a) Straight line distances between the nodes. (b) Travel times between the nodes.

is correct – if traffic is light, the true time might be less than 10, breaking the other rule. This does not mean that we cannot use the A* algorithm. It just means that the algorithm can no longer guarantee to find the best route.

There are also occasions when we need an algorithm to sacrifice the optimality of the route for the sake of speed. Imagine that you are using a satnav in your car and you miss a turning or there is a diversion. You need the system to recalculate the route as quickly as possible rather than taking a long time trying to find the absolute best route. The A* algorithm is an example of what is called a best-first search in that at each stage it tries to assess which of the various routes forward from the current node is most likely to be on the best route to the destination. However, because this is based on a heuristic, it will sometimes prove to be wrong and what seems like a good route will turn out to be not so good. The problem is that the algorithm may already do quite a lot of work before it discovers this. As an example, consider the network in Figure 11.26.

The best route from start to end is via node L with a distance of 20. However, because the route via node A sets off more directly towards the end, it will be selected by A*. This will also be true for nodes B, C, D and E all of which have f scores of less than 20. It is only when node F is reached that the f score will exceed 20 – in fact it will be 20.5, which is the distance from start to F (12) plus the h score for F, which is 8.5. At this point, the algorithm will

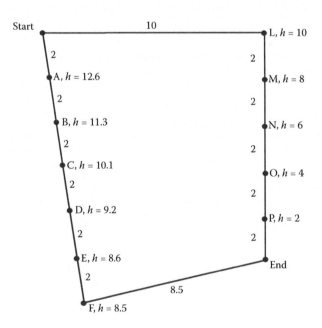

FIGURE 11.26
Imaginary road network which illustrates the advantage of ε-admissable heuristics.

choose L as the best candidate node and start processing all the nodes along the route from L to the end. However, the difference in total route lengths is slight. Would it not be better to stick with the original route and save time at the expense of a slightly less optimal route? This is the approach taken by a series of algorithms, which are called ε-admissable. We have seen that an admissible heuristic is one which never overestimates the cost of reaching the goal. ε-Admissable heuristics are ones which relax this rule slightly to speed up finding a solution. They do this by setting conditions for selecting the next node which give preference to carrying on along the current path, instead of switching back to an earlier one. However, there is a limit to how long they will persevere and this limit is that the final route should not be any more than $(1 + \varepsilon)$ times the length of the optimal route, where ε is the parameter used to modify the heuristic. One of the simplest is the weighted A* algorithm, which gives extra weight to the $h(n)$ factor in the calculation of the f score, that is

$$f(n) = g(n) + (1 + \varepsilon)h(n)$$

To see what effect this has Figure 11.27 shows the calculations of the f scores for nodes a to f and node l using the original h values and then using a weighted value in which $\varepsilon = 1$.

When the original heuristic is used, the f values along the route A–F are smaller than the value for L up until node F. At this point, the algorithm would select L and proceed down the other route. Because this is straight, the f values do not change and so the final route is start–L–M–N–O–P–finish with a total distance of 20. With a weighted heuristic with $\varepsilon = 1$, each h value

	g	h	f	$(1 + \varepsilon)h$	f
A	2	12.6	14.6	25.2	27.2
B	4	11.3	15.3	22.6	26.6
C	6	10.1	16.1	20.2	26.2
D	8	9.2	17.2	18.4	26.4
E	10	8.6	18.6	17.2	27.2
F	12	8.5	20.5	17.0	29.0
L	10	10.0	20.0	20.0	30.0
M	12	8.0	20.0	16.0	28.0
N	14	6.0	20.0	12.0	26.0
O	16	4.0	20.0	8.0	24.0
P	18	2.0	20.0	4.0	22.0

FIGURE 11.27
Calculation of f scores for nodes under distance heuristic (h) and ε-admissable heuristic with $\varepsilon = 1$.

is multiplied by 2. Because node F is nearer to the finish than node L, this gives node F an f score of 29 compared with 30 for L and so the route becomes start–A–B–C–D–E–F–finish with a total distance of 20.5. In this case, extending the route by 0.5 units has resulted in a halving of the number of nodes processed.

Further Reading

Gardiner (1982) provides a nice discussion on the importance of structuring network data, in the context of handling information on river networks. Haggett and Chorley (1969), although writing in a largely pre-computer age, cover many of the issues which arise in trying to solve problems using networks. The Dijkstra algorithm for shortest route is described by both Jones (1997) and Worboys and Duckham (2004). The A* algorithm was first introduced by Hart et al. (1968) and ever since a large number of variants have been developed. For a recent review, see Fu et al. (2006). Many practical implementations also preprocess the network data to make run times faster (e.g., Geisberger et al. 2012), which shows the important role that the data structure plays in algorithm efficiency. The relationship between road distance straight line distance was studied by Phibbs and Luft (1995) who were interested to see whether straight line distances were good enough to use in planning emergency service provision. The heap data structure will be covered by any computer science text such as Cormen et al. (1990).

12

Strategies for Efficient Data Access

We have now considered the data structures and algorithms which are used to store and process a range of different types of spatial data. In this chapter, we consider a range of strategies which are not specific to any particular data type, but are generally used to increase the efficiency of accessing large spatial datasets. To understand the factors which affect the speed of data access, let us return to a slightly modified version of the model of computer architecture which was introduced in Chapter 1, but this time focusing solely on data storage. In this model, data can be stored in one of four areas:

- The CPU
- Memory
- Local storage
- Networked storage

The majority of data processing takes place in the CPU and so a key issue is how to move data to and from the CPU as efficiently as possible. The CPU itself can only hold a few pieces of data at a time, so information is first held in memory and then passed to and from the CPU. However, there is also a limit on the size of memory, and memory is wiped clear when the computer is turned off and so data have to be placed on some form of storage device for permanent storage. This could take many forms, but for simplicity it will be assumed that this is a disk drive. Increasingly, the computer will be connected to the Internet and can therefore access data which are stored on disks that are connected to other computers on the network. As you go down the hierarchy from the CPU to the network, the amount of data which can be stored increases, but the speed with which it can be accessed decreases. It is difficult to be precise about the relative speeds, but there is probably at least an order of magnitude difference between each level.

Many spatial datasets are too large to be held in memory and so retrieving data from them will involve making a transfer of data from disk to memory or even from a remote computer to a local disk and then to memory. As these transfers are all potentially slow, a considerable amount of thought has gone into developing methods to make them as efficient as possible, and this has involved three sets of techniques:

- Creating indexes which allow the correct data to be found on the disk quickly

- Storing data on the disk in a way which means that it can be transferred to memory as quickly as possible
- Keeping a local copy of the data in case it is needed again, a process known as caching

To understand the principles at work, it is useful to begin with the real-world analogy of undertaking research in a library. Let us assume that you wish to follow up one of the topics in this book and you decide to start by consulting some general textbooks on computer science. Your first step will be to consult the library catalogue to see which relevant books the library has and where they are shelved. Then, you will need to go and retrieve the books from the shelf. This will probably take longer than searching the index because it involves physically going to the shelf, searching for the correct books and bringing them back to your desk. Once you have finished looking at a book, you would not return it immediately because you might decide you need it again.

Keeping the books for a while is very analogous to caching, and does not really require further explanation for the moment. Having gone to the effort of fetching a book, it makes sense to hang on to it for a while rather than returning it to the shelves immediately.

Searching the catalogue is a good analogy for using a database index, and fetching the books is similarly a good analogy for retrieving data from a disk. Let us consider each stage in turn. How well the catalogue has been designed will clearly affect how quickly you can find the information you need. In a modern system, the catalogue will be a database and will probably allow you to search by author name, title, topic and date. In an older library, the catalogue would have been on index cards and probably organised by author surname. If there was not a version of the cards sorted by topic, you would have had to rely on the fact that in most libraries non-fiction is stored according to a classification of the subject of the book. Therefore, you would need to find the classification number for general computer science and discover where books of this classification were shelved. So, clearly the way the books are indexed is going to affect how quickly you can find what you want. We have already seen that physically fetching the books is likely to be a slow process and retrieving data from disk storage is also slow for similar reasons. Disks hold the data on a circular magnetic disk, with the data stored in a series of sectors arranged in concentric circles. Retrieving data involves finding out which sector it is stored in, moving the disk read head to a position over this sector and then waiting for the start of the sector to come round as the disk spins. All this happens very quickly of course, but compared with the transmission of data from memory to CPU, which is a purely electronic process, the transfer of data from disk is considerably slower. Because of this, it matters how the data are stored on disk and how much is retrieved at one go. Let us return to the library analogy for a moment. In storing books in a library, one of the issues is how best to make use of the available storage space. Assume for the moment that this was the only consideration. How could the space be used most efficiently? One way would be to group books purely on the basis of their height, and have a series of shelves of increasing height. Each new book would then be shelved in the next available slot on the smallest height shelf which would accommodate it. As long as the catalogue was updated correctly, such a library would still be usable. However, books on similar topics or by the same author will no longer be shelved together so that if there are 10 computer science textbooks; retrieving them will involve 10 trips to the shelves and back. Again, this is exactly analogous to retrieving data from a computer disk, and a key to the design of a data storage strategy is try and ensure that records which are likely to be retrieved together are stored close to each other on disk. There is a particular reason why this is important on disks and this is where the library analogy breaks down. Data are always returned from a disk in chunks of a fixed size which are variously called pages or blocks. For instance, in many versions of the Windows operating system, the standard page size is 4 kB or 4096 bytes. So, whether a program requests 1 byte or 4000 bytes, the operating system will retrieve the 4 kB page which contains the necessary data. It is as if you had a library in which you could not take just one book, but were forced to take all the books from one section of the shelves at each visit. The reason things are done this way is again part of speeding up data access since it is much quicker to return data in fixed-size pages than spend time trying to locate the exact parts of each page which are needed. This is done at a later stage once the data are in memory, and everything can be done more quickly. As data are always returned in complete pages, it is far more efficient if data items which are likely to be used together are stored together on the disk, and not scattered around.

It follows from this that making database queries run quickly involves having an index which can find out where in the database the appropriate records are, and then having a data structure which clusters those records together on the disk. This is true for all databases, not just GIS, but spatial data poses a particular problem because of its 2D nature. Indexing and clustering are both made easier if there is a natural sequence in the values. The most obvious examples in everyday life are the index to a book, in which the entries are in alphabetical order, and a telephone directory where the entries themselves are in alphabetic order of surname. However, spatial data do not possess such a simple sequence. Consider the data in Figure 12.1.

The figure shows simple forms of the two main spatial data types – vector points and raster pixels. A typical vector query might be to find the point which is nearest to point 1. If we could sort the points according to their spatial location, then the nearest points to 1 would lie on either side of 1 in this list. The problem is that there is no way to produce such a list. It is possible to calculate the distance between point 1 and each other point and this will produce a list which could be sorted to find the answer to our question. But, this is not a general list. We cannot use it to find the nearest point to point 2, for example. The problem is no easier in raster. Consider the pixel which

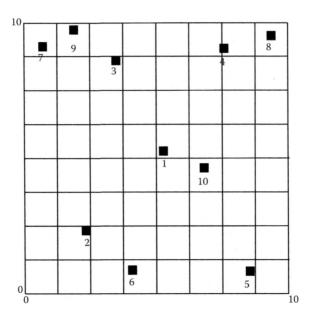

FIGURE 12.1
Vector and raster data.

contains point 1. We might wish to find the values in the neighbouring pixels. In a list sorted by location, these pixels would be next to 1. However, even if we just take the row and column neighbours, and ignore the diagonals, we have four, and all four cannot be next to the entry for pixel 1.

Two approaches to solving this problem have been taken. The first is to devise data structures specifically for dealing with 2D data. The second is to devise indices which combine the two dimensions into one value which can be treated as 1D data and can therefore take advantage of data structures and techniques developed for non-spatial data. Both approaches make use of tree data structures for indexing the data and so we will consider some general issues about these data structures first. Next, we will look at a range of different spatial indexes and finally end the chapter by considering the issue of caching.

12.1 Tree Data Structures

Whether spatial data are treated using both X and Y coordinates, or by combining these into a single value, tree data structures are widely used in indexing and accessing it. We have encountered tree data structures twice before – the quadtree, as a data structure for raster data (Chapter 7) and the binary heap, as a data structure for speeding up the shortest path algorithm

Random	4	7	18	10	1	3	11	14
Numerical	1	3	4	7	10	11	14	18

FIGURE 12.2
Operation of binary search. Upper table: 8 numbers in random order. Lower table: same numbers in numerical order.

(Chapter 11). However, trees have a much wider range of uses than this, and are in fact one of the commonest data structures used in computing.

To explain the principle behind the use of trees for searching, let us start with a very simple query of finding an individual item in an index. This is equivalent to searching the theatre database for a particular play (Chapter 2) or finding a particular book in a library catalogue. To simplify the explanation, we will use a small set of whole numbers and our query will be whether 3 is one of these numbers. Figure 12.2 shows the eight values arranged both in a random sequence and in numerical order.

If the numbers are not in order, we are going to have to look at each one in turn. If the number we are after is not among them, we will look at all eight numbers. If it is, we will look at half of them on average before finding it. This search is therefore an $O(n)$ algorithm. However, if the numbers are sorted, as in the lower half of Figure 12.2, we can perform what is called a binary search. This was described in Chapter 6, but is worth summarising again since the logic underlies much that is to follow. We begin by examining the middle number in the list. If it is what we are looking for, we have finished. If it is not, the number we want, and it is greater than 3, we repeat the process on the first half of the list. If it is less, we examine the second half of the list. We can only divide the list into half three times before we either find the item we want or conclude that it is not in the list. We now have an $O(\log n)$ algorithm. For a small list like this, the difference is small; but, as n grows, $O(\log n)$ becomes much better than $O(n)$.

The pseudocode for a procedure which implements a binary search is as follows:

```
1.   procedure binary_search(k)
2.   Array SORTED[1..n]
3.   first = 1
4.   last = n
5.   found = FALSE
6.   repeat until first == last or found == TRUE
7.       middle = ((last - (first - 1))/2) + first
8.       if (SORTED[middle] == k]
9.           found = TRUE
10.      else if (SORTED[middle] < k)
11.          first = middle + 1
12.      else
13.          last = middle - 1
```

```
14. if found == TRUE or SORTED[first] == k
15.     return middle
16. else
17.     return 0
```

The procedure keeps two variables, first and last, to keep track of which part of the sorted array to deal with. Initially, these point to the start and end of the array. Line 7 calculates the middle position, which in this case will be 5. The fifth element is 10, which is greater than 3. This causes the variable last to be set to the array element to the left of middle; thus, the second time, through the repeat loop, elements 1–4 will be searched. The middle in this case is entry number 3 which has a value of 4. Next time elements 1–3 will be searched and the middle one is the second element. This contains the search term and so the procedure will return a value of 2.

If the repeat loop continues until the first and the last become equal, this means that no further subdivision is possible and the repeat loop ends. A final check is necessary to see whether this final element of the array contains the value we are looking for. If not, then a value of 0 is returned to indicate that the search item is not in the array.

We can also use a tree data structure to simplify this procedure and avoid the constant calculation of which element of SORTED to look at next. These same eight numbers are shown stored in what is called a binary search tree in Figure 12.3. Like the quadtree, the tree is made up of two types of nodes. Leaf nodes are those which have a number stored in them, but no further numbers below them. Branch nodes also store a number, but in addition, they point to two further branches of the tree – one on the left and one on the right. Every branch is organised so that all the nodes below it on the left contain numbers which are smaller than it, and all the ones on the right are larger.

Check for yourself that this is so in the case of Figure 12.3. The root node contains the number 10. Every number down the left-hand branch is less than

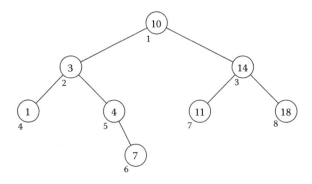

FIGURE 12.3
Binary search tree. Numbers inside circles are values held in nodes. Numbers next to circles are node IDs.

10. Every number down the right-hand branch is greater than 10. The same is true of the node containing the number 3 – this has 1 on the left and 5 on the right. Figure 12.4 shows how this tree might actually be stored in an array.

The array has five columns and eight rows. Each row contains the following five items for a node:

1. Value stored in the node
2. Pointer to the left child
3. Pointer to the right child
4. Pointer to the parent
5. Pointer to location of data

Note that the node IDs are not stored explicitly because they are not needed as the row number functions as an ID. For example to find the left child of node 1:

```
Array TREE[1..8,1..5]
Left_pointer = TREE[1,2]
Left_value = TREE[left_pointer,1]
```

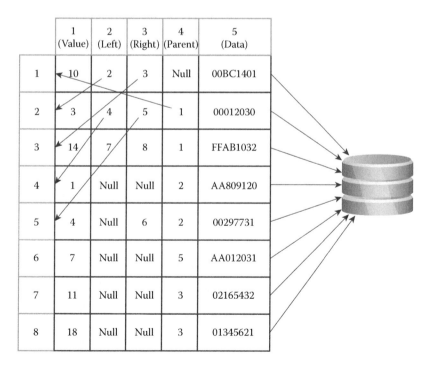

FIGURE 12.4
Binary search tree stored as a linked list in an array. The array elements are outlined in bold, the row and column numbers in grey. Pointers for nodes 1 and 2 are shown with arrows.

In fact, there is no need to put the pointer into the left pointer variable – you could just use

```
Left_value(i) = TREE(TREE(i,2),1)
```

where i is the number of the node whose left child you are interested in.

Node 1 (the root) points to nodes 2 and 3. As it is the root, it does not have a parent, so this pointer is stored as 'null'. Node 2 has pointers to nodes 4 and 5 as its children and 1 as its parent. Node 4 is a leaf node, and so has the pointers to children set to null. The final column is shown as containing an address of where on disk the data for that node is stored. The exact way in which locations on hard disk are handled is beyond the scope of this book. However, the addresses are simply numbers which tell the system which cylinder and sector to go to on the disk. Here, they are shown as purely imaginary 32-bit words, but represented in hexadecimal, so that each character in the number represents 4 bits. We will come back to the issue of storing the data which are being indexed later – for the moment, we will just focus on the way in which the tree data structure is used.

The procedure for performing a binary search using a tree has the following overall structure:

```
1. Procedure search(node)
2. Look at value of node
3. If node is LEAF STOP: 'Not found'
3. If same as search value STOP: 'Found'
4. If search_value <current search(left-child)
5. Else                       search(right-child)
```

This is a recursive procedure. Starting at the root of the tree, the procedure checks to see whether the current node is what we are looking for. Then, as long as the current node has children, it makes a call to itself to repeat the process using the appropriate child node. In detail, the algorithm might be implemented as follows:

```
1.   search_bst(T,k)
2.   if T == NIL then
3.       print 'Not found'
4.   else if k == value[T]
5.       print 'Found'
6.   else
7.       if k < value[T] then
8.           search_bst(left[T],k)
9.   else
10.          search_bst(right[T],k)
11. return
```

The procedure is passed a pointer T to the root of the tree. When the procedure is first called, T should not have a value of NIL so line 4 will be executed.

This tests whether the value at the current node matches the search item, *k*. If it does, the search has been successful. If not, search_bst is called again, this time with a pointer to the right or left subtree from the current node. Assume that this procedure is called to see whether 3 is included in the list. The sequence of calls to SEARCH_BST would be

```
1. search_bst(1,3)
2. 3 < 11
3. search_bst(2,3)
4. 3 = 3
5. Found
```

On line 5, the second call to search_bst succeeds. At this point, the procedure returns to the procedure which called it. This was actually the previous call of search_bst, which at the time was on line 8. This call of the procedure then returns to the program which called it, terminating the process.

The binary search tree shown in Figure 12.3 is well balanced, which means that the majority of branch nodes have two children. This is equivalent to dividing the original sorted list into half at each stage of the search, and so the search process will be O(log *n*) in this case. However, it depends on how the tree structure is built from the original data as to whether this is true. To take an extreme case, imagine taking the sorted list of eight numbers, and adding them to the tree one at a time in sorted order. Every node would have a single, right-hand child node because each successive number added would be larger than the previous one. The list would therefore be eight levels deep, and no better for searching than the original, brute force search through the sorted list. This is an extreme case but to avoid it might be worth sorting the original list of numbers, which can be done on O(*n* log *n*) time, since this makes it easy to build a balanced tree. The root of the tree is the middle value in the sorted list, since by definition this has equal numbers of points on either side of it. If we apply the same logic to each half of the list at each subdivision, we will create a balanced tree.

To see how the binary tree can be applied to spatial data, we will begin by considering a 1D case. Figure 12.5 shows the points from Figure 12.1, but with their *Y* coordinates all set to the same value so that they can be plotted in order of their *X* coordinate as shown. The problem is to identify which of them lie within the range 3.0–6.0 as shown by the dotted lines. This is now a spatial query and in fact it is an example of what is called a range query – finding all the objects which fall within a given spatial window. Clearly, the brute force approach is to compare the *X* coordinate of each point with the *X* coordinates of the ends of the range. However, by utilising a search tree, we can reduce the number of points for which we have to make this comparison. A standard binary search tree can be used to store the *X* coordinates of the points as shown in Figure 12.5. This has been constructed in exactly the same way as the earlier example, but using the *X* coordinates of the points rather

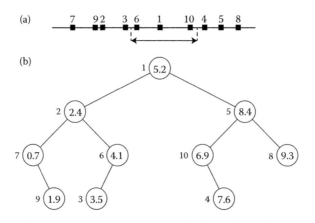

FIGURE 12.5
(a) Ten points ordered by their X coordinate along a single line. (b) Binary search tree which is based on the X coordinates of the points. The coordinates are shown inside the nodes and the point ID to the left.

than their IDs. Hence, all the points on the left-hand side of the tree are to the left of the root, which is point 1.

To find out which points fall within the range, we start at the root, which is point 1, and test whether this is inside the range. It is, which means that points to both left and right may also be inside the range so we must examine both the nodes below the root. We begin with the left child which contains point 9. This is to the left of the left-hand end of the range. This means which we do not need to investigate the tree below the left-hand child of 9 because none of the points in this tree can be to the right of point 9. We do have to examine the right-hand child, which is point 3. This is also to the left of the range, so we only have to examine its right-hand child – point 6. When we look at the right-hand branch of the tree below the root, we find that since point 4 is to the right of the range, we only need to consider its left-hand child, point 10 – points 5 and 8 can safely be ignored. The algorithm for this search is the same as a basic binary search, except that instead of looking for an exact match with the search key, we need to look for whether the X coordinate of the candidate point falls within the range. In this simple example, of the 10 points, we only need to inspect five of them to discover the three which actually fall in the range.

12.2 Indexing and Storing 2D Data Using Both Coordinates

Tree data structures are a very efficient way of indexing 1D data and so in indexing 2D data, it clearly makes sense to try and develop tree data structures for 2D data. First, we will consider two methods which consider both

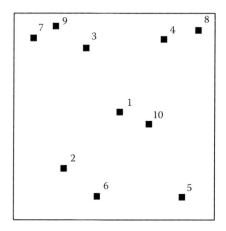

FIGURE 12.6
Vector points.

the X and Y coordinates of the data and we will see how they can be applied to our original vector points as shown in Figure 12.6.

The first approach is based on extending the binary tree to extra dimensions to create what is called a k-d tree. k originally referred to the number of dimensions being handled, since the basic idea was not limited to two dimensions, but could apply to three or even higher numbers of dimensions. In theory then one could talk about a 2-d tree, but the term k-d tree has become generally accepted for referring to this data structure.

The k-d tree simply extends the logic of the binary search to extra dimensions by considering the X and Y coordinates alternately. Let us assume that we begin with the X coordinate and so we start the tree by finding the 'middle' point when the points are sorted into X coordinate order. Middle is in quotes because there are several methods which could be used to determine which point this should be. We will use the median point, which in this case could be either 1 or 6 and so we will pick point 1 at random. This splits the remaining points into two sets:

```
Left of 1: {7,9,2,3,6}
Right of 1: {10,4,5,8}
```

Each of these is considered separately and to decide what the two child nodes should be below 1, we sort each of these sets on the basis of their Y coordinate and again pick the middle one.

```
Left of 1: {6,2,3,7,9} Median = 3
Right of 1: {5,10,4,8} Median = 10 or 4. 10 is selected at random.
```

We now have our two child nodes − 3 and 10, but we need to decide which of them is the left child and which is the right. The rule is that the one with the lower coordinate value goes on the left. Since node 1 was selected on the basis of its X coordinate, its children are allocated using the same coordinate, so that point 3 becomes the left-hand child of point 1 because it was the median point of all those which were on the left of point 1.

The same procedure is followed for all the other points. To insert a point, we start at the node of the tree and proceed down until we reach the first available position which is not already occupied. At the root level, we decide which branch to look down by considering the X coordinate of the new point and the point currently in the tree. After this, at each level down the tree, we alternate between considering the Y and X coordinates.

The range query for the k-d tree is then essentially the same as the procedure for searching a binary search tree:

```
1.   procedure range_search (T,RANGE,level)
2.   if T == NIL return
3.   if T inside RANGE Mark T
4.   if level is even then
5.        if (T[x] > RANGE[x1]) then
6.             call range_search(T[left],RANGE,odd)
7.        if (T[x] < RANGE[x2]) then
8.             call range_search(T[right],RANGE,odd)
9.   else
10.       if (T[y] > RANGE[y1]) then
11.            call range_search(T[left],RANGE,even)
12.       if (T[y] < RANGE[y2]) then
13.            call range_search(T[right],RANGE,even)
14.  return
```

Now that we have a 2D search, the range is a rectangle, defined by the X and Y coordinates of its lower left and upper right corners as shown in Figure 12.7. When range_search is first called in this case, it is passed a pointer to the first node in the tree, which is point 1. This is not inside the range, so it is not marked. On this first call, the variable called level is set to the value even. The X value of the current root is then compared with the X values of the range window, as was done for the 1D case. In the case of the tree in Figure 12.7, the test on line 6 will fail, which means that none of the branches to the left of point 1 will be tested. On line 8, point 1 is to the left of x2, so range_search is called to perform a search starting with the right-hand child of point 1. Note that this time the level variable will be passed as 'odd', so that it is the Y coordinates which will be considered when looking at point 10.

When point 10 is checked, it will be found to be in the range window, so it is marked. Since it is in the window, both its left and right children will have to be examined. When point 5 is checked, it is not in the window. However, because the X value of point 5 is within the range of X values of the window, both its children will be searched. Point 5 does not have a right-hand child,

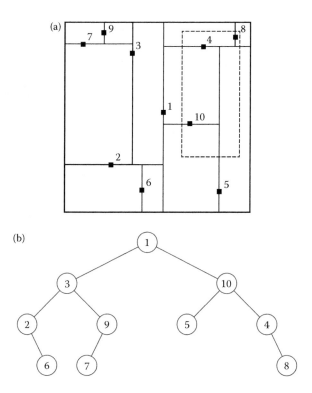

FIGURE 12.7
k-d tree. (a) Map of the points. The search window for the range search example is shown with a dashed line. (b) Tree structure after all 10 points have been inserted.

so this call to range_search will terminate immediately. However, point 4 will be found to be within the range and will be marked as will its child node, point 8.

As with the 1D case, the use of the tree structure, sorted according to the *x* and *y* coordinates, has reduced the number of points which need to be searched. In this example, of the 10 points, only 5 have been considered, of which 3 are in the window. A brute force algorithm for range searching would compare the *x* and *y* coordinates of each point with the coordinates of the window and would be an O(*n*) operation.

The *k*-d tree has an expected depth of O(log *n*) as with many binary structures. However, this is not the efficiency of the search algorithm. Consider two extreme cases. If the search window happened to contain all the points, then the entire tree would have to be searched, so search_range would be called *n* times. The other extreme is if none of the points is in the search window. In this case range_search would only ever have to consider one branch of the tree at each level, so would run log *n* times at most. Clearly, the efficiency therefore depends both on the number of points in the tree and on the number which falls in the window. de Berg et al. (1997) show that the efficiency is actually $O(\sqrt{n} + k)$, where k is the number of points inside the search window.

A second approach to indexing 2D data is based on the idea of the MBR which was introduced in Chapter 4. We saw earlier that a range search on 1D which involved looking at the X coordinates of the data points and comparing them with the X values of the range criteria. The natural way to extend this idea to 2D is to use a rectangle to define the search and MBRs to describe the location of the objects being searched. This idea leads to what is called a R-tree which is illustrated in Figure 12.8.

The points have now been grouped together spatially and each group has been defined by its MBR. The grouping is done hierarchically. Initially, the 10 points are put into two groups as follows:

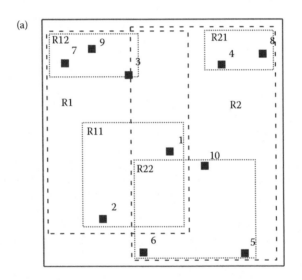

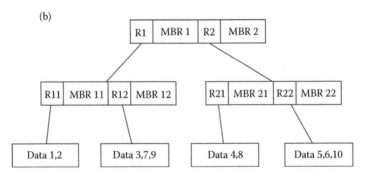

FIGURE 12.8
R-tree. (a) Subdivision of the area by rectangles. Dashed lines indicate rectangles at first level of tree and dotted those at the second level. (b) Tree structure for storing rectangles.

```
Group 1: {1,2,3,7,9}
Group 2: {4,5,6,8,10}
```

At the second level, group 1 is further split as follows:

```
Group 11: {1,2}
Group 12: {3,7,9}
```

Figure 12.8b shows the tree structure which is used to store the information about the rectangles and the points they enclose. The root of the tree contains pointers to the two first level rectangles, R1 and R2. The only information stored in the tree is the MBR of each rectangle and a pointer to the node which contains the information about each rectangle. This structure is repeated at the next level down, so that the entry for node R1 contains the MBRs for its two children R11 and R12. However in this case there are no further subdivisions and so the node contains pointers to where the data points for R11 and R12 are stored.

The *k*-d tree was a structure which created a tree in which every data point was represented by a single node. The R-tree differs in that the rectangles in the tree can contain several points instead of just one. If the points inside each rectangle in the tree are stored together on disk, this will lead to improved clustering of data on the disk. In addition, the R-tree is not restricted to storing points. As Figure 12.8 shows, the leaf nodes contain the MBR of each of the objects which fit inside the final rectangle and there is no reason why the objects cannot be lines or polygons. In fact, the *k*-d tree can also be used to index lines or polygons. The basic tree as described above could be used for this purpose by using the centroid of the object as a point in the tree. Alternatively, there are other data structures based on the idea of successively dividing space in the X and Y planes, which are designed to handle 2D objects.

With any of the indexing structures being described in this chapter, there is also the issue of where the actual data should be stored. With the *k*-d tree (Figure 12.7) each node in the tree represents a data point; whereas with the R-tree (Figure 12.8), the non-leaf nodes simply contain the pointers to elements lower down the tree, and the data points are all held in leaf nodes. Storing the data itself in the *k*-d tree could make the data structure complicated and difficult to search, whereas this would not be an issue with the R tree. However, the more the data are stored in the tree, the larger the index becomes and it may become so large that it will not fit inside memory. The issues of paging data from disk which apply to the spatial data then also become relevant for the index. In fact, we will see later that one of the commonest index structures, the B-tree, is designed specifically to deal with indexes which are too large to fit in memory. It should be clear from this brief discussion that the design of efficient indexing methods for spatial data is a complex process. With this in mind, let us turn to indexes based on converting 2D locations into 1D data.

12.3 Space-Filling Curves for Spatial Data

The second approach to dealing with the fact that spatial data have two coordinates is to try and combine these two numbers into a single number, which in some way reflects the positions of objects in 2D space. We saw in Chapter 7 that in building a quadtree, interleaving the bits in the X and Y coordinates of raster pixels produced the Morton code, which could be used to locate the nodes of the tree. This idea of combining the X and Y coordinates into a single number has a much broader application than quadtrees and can be used in numerous ways to index and store a wide range of spatial data.

When the Morton codes for a series of pixels are joined in sequence, they form the pattern shown in Figure 12.9a–c. If we regard the squares in this figure not as pixels, but as regular subdivisions of space, then we can see that the Morton code for each one is a way of recording its location as a single number. As Figure 12.9a–c shows, the sequence of the Morton codes crosses space in such a way that points which are next to each other in Morton sequence are also close in space. What is more, if we produce smaller subdivisions by halving the sides of the areas, the Morton pattern for the new areas is simply a smaller version of the original.

The pattern produced by connecting the Morton codes in order is one example of what is called a space-filling curve (SFC). The word curve may seem an odd one to apply to such an angular line, but to a mathematician, a curve is simply a line which is not straight. The curve produced by the Morton code is also called the Z-curve and is just one of many such curves. The Hilbert curve, which is shown in Figure 12.9d–f, is another which is widely used in GIS applications.

Space-filling curves are always recursive in nature, which means that by the application of a simple rule, it is always possible to produce a version of the curve at finer and finer spatial resolutions, as shown by the diagrams in Figure 12.9. This process can be repeated infinitely and eventually the curve will fill the whole of the space. This is not really important from the perspective of GIS, but these curves share two properties which make them very useful. First, it is possible to calculate the position along the curve of any point from its X and Y coordinates. We have already seen how this is done for the Morton code. It is also possible for the Hilbert curve although it is rather more complicated. The second property is that these curves all produce sequences in which points that are neighbours along the curve are also close in 2D space. These two properties mean that space-filling curves can be used in two ways in improving access to spatial data:

1. By storing data on disk in order according to an SFC, clustering of spatial neighbours on disk is improved.

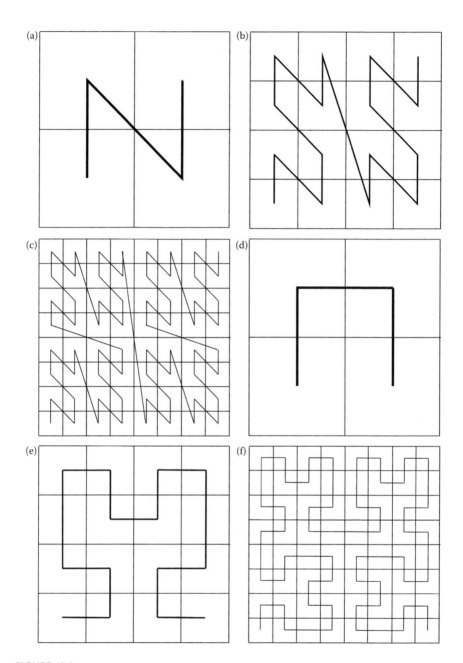

FIGURE 12.9
Space-filling curves. (a)–(c) Morton or Z-curve. (d)–(f) Hilbert curve.

2. By using the code produced to index data on disk, a range of indexing methods which have been developed for non-spatial data can be
applied to spatial data.

We will consider each in turn.

12.4 Spatial Filling Curves and Data Clustering

First, we will see just how good an SFC is at keeping spatial neighbours
together and then we will explore how the codes these curves produce can
be used to index and store spatial data.

To see how the Morton code can produce an improved spatial sorting, let
us consider the same points we looked at earlier. In Figure 12.10, they are
shown with the regular grid and the Morton curve superimposed.

Now imagine that we stored these points in two lists, one sorted by X coordinate and one by Morton order as shown in Figure 12.11. For each sequence, we
can calculate the geographical separation between points that are next to each
other on the list and we can see that this is lower on average for Morton order.

It is also possible to analyse how these orders might perform in answering
a query. To do this, we will use a simpler version of the tests which have been
done in the literature. In assessing these sort orders, what is of importance is
how many disk accesses it will take on average to retrieve the data for a spatial
query. With 1D data, such as a list of names, it should be possible to sort the

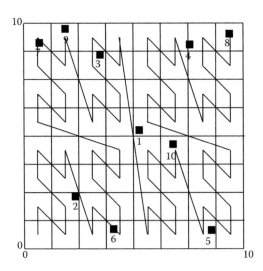

FIGURE 12.10
Morton curve with points superimposed.

Sequence	Points										Average distance to nearest neighbour
X order	7	9	2	3	6	1	10	4	5	8	5.6
Morton order	2	6	7	9	3	1	10	5	4	8	3.9

FIGURE 12.11
Points stored in the X coordinate and Morton order and average geographical separation between successive points.

values so that all the entries which satisfy the query can be retrieved with the minimal number of disk accesses. For instance, imagine the query is for all names beginning with S, there are 3000 of them and that each disk access can retrieve 1000 names. As long as the S entries are stored together on disk, this means three disk accesses. However, if they are scattered across the disk randomly, this will increase the number of accesses and in theory it could mean as many as 3000 accesses, one per name. What is important therefore is what is called clustering – the degree to which related entries are stored together.

The first test is a version of some work which was done by Mark (1989). Figure 12.12 shows 16 square areas indexed by row order, Morton order and Hilbert order. Assume that the data for these areas are stored sequentially on disk in the numerical order shown. How efficiently could we access the row and column neighbours of any given square? To look at this, we calculate the number of data items we would need to retrieve. For instance, the square labelled 0 in all three figures has two neighbours – the square to the right and the one above. To access the data for the square above 0 for each sort order, we would need to retrieve the following values:

```
Row order: 1,2,3,4 = 4 data values
Morton order: 1 = 1 data value
Hilbert order: 0,1,2,3 = 4 data values
```

To retrieve the square to the right we would need 1, 2 and 1. Averaging the values gives us an overall cost of

```
Row order: Cost = 2.5
Morton order: Cost = 1.5
Hilbert order: Cost = 2.0
```

If we do the same calculation for all neighbours for all squares, we get the results shown in Figure 12.13.

Two summary statistics are shown for each curve. The mean value is a measure of how many numbers would have to be retrieved on average to be sure of getting all the immediate neighbours of a square. The median is a measure of how many numbers would be needed for the best 50% of cases. The results are somewhat mixed. The mean distance to neighbours is actually worse for the Hilbert curve than for row order, and the Morton order is only slightly better. However, the Hilbert order is the best when measured

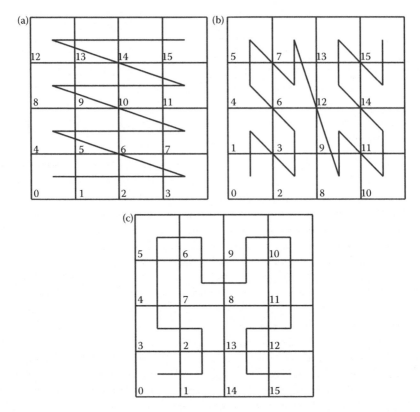

FIGURE 12.12
Sequence of points in three curves. (a) Row order. (b) Morton order or Z-curve. (c) Hilbert curve.

using the median. This gives us a first clue that the answer to which of these sort orders is best may not be straightforward. Let us look at another test.

The tests that are usually performed consider how efficient sort orders are at returning range queries and Figure 12.14 shows the results of a series of simple rectangular range queries run on the figures in Figure 12.2.

2.5	2.0	2.0	2.5		1.5	3.0	3.0	1.5		1.0	1.7	1.7	1.0
3.0	2.5	2.5	3.0		2.0	3.0	3.0	2.0		1.7	2.5	2.5	1.7
3.0	2.5	2.5	3.0		2.0	3.0	3.0	2.0		1.7	4.5	4.5	1.7
2.5	3.0	3.0	2.5		1.5	4.5	4.5	1.5		2.0	7.5	7.5	2.0

Row order	Morton order	Hilbert order
Mean: 2.63	Mean: 2.56	Mean: 2.81
Median: 2.50	Median: 2.50	Median: 1.83

FIGURE 12.13
The cost of retrieving the rook neighbours of cells in Figure 12.12.

Column	Row		Morton		Hilbert	
x row	Median	Mean	Median	Mean	Median	Mean
4 × 1	3.0	3.0	10.0	10.0	9.0	9.5
4 × 2	7.0	7.0	11.0	11.7	11.0	11.0
2 × 2	5.0	5.0	5.0	5.0	3.0	5.4
2 × 4	13.0	13.0	7.0	8.3	7.0	9.0
1 × 4	12.0	12.0	5.0	5.0	5.5	5.5
Mean	8.0	8.0	7.6	8.0	7.1	8.1

FIGURE 12.14
Summary statistics for a series of range searches on the grids in Figure 12.12.

The first column shows the query which is being run. For example, 4×1 means retrieving an area of four columns and 1 row. Four such queries are possible with the data in Figure 12.2; for each one, we calculate the along the sort order, between the first and last squares retrieved. For example, the results for retrieving the four values in the uppermost row in Figure 12.2 are as follows:

```
Row order: 12-15, Range = 3
Morton order: 5-15, Range = 12
Hilbert order: 5-10, Range = 5
```

Doing this for all four rows and taking the mean and median of the ranges produces the values in the first row of Figure 12.14.

The queries include retrieving each row (4×1), retrieving each column (1×4), retrieving square windows (2×2) and retrieving rectangular windows (4×2, 2×4). The results show that for any particular query, each sort order can perform best and that this depends partly on the measure used – mean or median. If whole rows are retrieved, row order is best on both measures, but it is by far the worst for retrieving columns. Morton and Hilbert orders do better for windows which contain more columns than rows, but this is because of the way the sequence has been done in both cases: rotate each through $90°$ and this result would change.

However, these simple tests do confirm what is generally found in the literature, which is that over a wide range of queries space-filling curves tend to lead to better data clustering and hence more efficient queries.

12.5 Space-Filling Curves for Indexing Spatial Data

The codes produced by space-filling curves can be used as the key to a binary search in exactly the same way as we did with the X coordinates

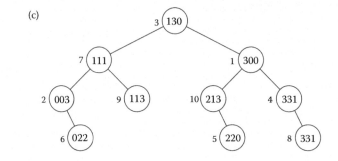

FIGURE 12.15
Points stored in a binary search tree according to their Morton code. (a) Points and the Morton codes of all cells. (b) Morton code for each point. (c) Binary tree.

earlier in the chapter. The conversion between the XY coordinates and the code is simpler for the Morton code than for the Hilbert curve, so this will be used as the example. Figure 12.15 shows the data points we used earlier in the chapter overlaid on an 8×8 grid of squares. The points have coordinates in the range 0.0–10.0. The squares have X and Y values which are integers in the range 0–7. Converting the X and Y values of each point to this 0–7 range using this formula

```
Integer coordinate = abs(decimal coordinate/10)
```

will produce a pair of integer numbers, which in binary will run between 000 and 111. Interleaving the bits, as described in Chapter 7, produces numbers in the range 000000 and 111111. To keep these Morton codes short, they are written in Base4 in Figure 12.15, and so they run from 000 to 333. Applying these calculations to each of our points produces the Morton code for each as shown in Figure 12.15b.

The Morton codes can then be placed in a binary search tree and used in a range query in exactly the same way as was done for the X coordinates of these points earlier in this chapter (Figure 12.15c).

One of the properties of the Morton curve is that for any rectangular area, the Morton code of the lower left corner will be smaller than of the upper right (or the same if the rectangle is small relative to the Morton grid). This means that answering a range query becomes a matter of finding the Morton code of the corners of the search window and retrieving all the points in the list with codes between these values. This could be implemented very simply in SQL for instance

```
SELECT ID
FROM POINTS
WHERE MORTON > MORTONLL AND MORTON < MORTONUR
```

Let us see how this would work for the rectangular area shown in Figure 12.15. The Morton codes for the query area are

```
Lower left: 0211
Upper right: 0331
```

If you look at the table in Figure 12.15, you will see that points 10, 5, 1 and 4 will be returned by this query. Three of these lie inside the query window, but because of the way the Morton curve criss-crosses the area, point 10, which is outside the window, is also returned. This means that some further action will be needed to check which of the returned points are actually inside the window. However, this further check is only needed for points returned by the range search and so compared with doing a range query by checking every point against the query window, we are saving processing.

To describe the range query in detail, we will use a different version of the binary tree as shown in Figure 12.16.

This stores the same set of Morton codes as Figure 12.15, but using what is called a balanced binary tree. In a balanced tree, the total number of nodes in the left- and right-hand halves of the tree is approximately equal. The same is true for any subtree, as far as possible. Balancing a tree places a limit on how deep the tree can become and this in turn makes queries more efficient on average because there are fewer levels in the tree to search. The tree in Figure 12.16 also stores the data values (or the pointers to where data are stored) in the leaf nodes, shown on the diagram as square nodes. The non-leaf nodes simply contain the index values which allow the tree to be searched and the

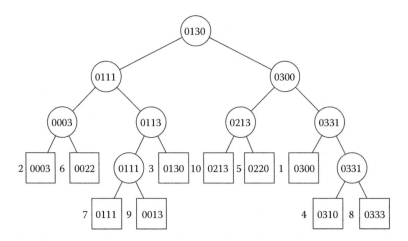

FIGURE 12.16
Balanced binary search tree. Leaf nodes are indicated by squares, non-leaf nodes by circles.

correct leaf nodes to be found, which is why the Morton codes appear two or more times. However, the tree still obeys the binary search tree property of right nodes being larger than the parent and left nodes smaller, which means that the leaves are stored in the Morton code order running left to right. The nodes themselves would contain the information shown in Figure 12.17.

Both have a relatively simple structure. Non-leaf nodes need contain only the Morton code and pointers to the left and right child nodes. Leaf nodes will have the Morton code and either the data for the object in question or a pointer to where on disk it is stored.

Let us see how a range search will work using this balanced tree, based on an algorithm given by de Berg et al. (1997). The range is defined by two values – range low (RL) and range high (RH). The algorithm consists of three elements:

1. Find the leaf nodes representing the low and high values of the range (or the leaves which have the closest Morton code and which lie inside the range) – these are referred to as leaf low (LL) and leaf high (LH).

ID	Morton code	Left child	Right child

Non-leaf node

ID	Morton code	Data

Leaf node

FIGURE 12.17
Contents of nodes in a balanced binary tree.

2. Find the point at which the path down the tree to these nodes splits into two.

3. Follow the left-hand branch and report all leaf nodes to the right of the path. Follow the right-hand branch and report all leaf nodes to the left.

To illustrate the principle, Figure 12.18 shows a full, four-level binary tree and the results of a range search between the values RL and RH.

Remember that in a balanced tree, the range search will always end in two leaf nodes and these are shown in grey and labelled as LL and LH. If you trace the path back up the tree from these leaf nodes, there will be a node in the tree where these paths meet. This will be the highest node in the tree which falls inside the range and so anything higher up the tree, or connected to a higher node can be ignored. The node where the paths split is labelled S and marked in dark grey in Figure 12.18 for the range search shown – you might like to try and find the split node for other pairs of leaf nodes – note that in many cases the split node will be the root of the tree itself. Starting from this split node, there will be a left child and a right child. As we trace down from each of these, when we find a non-leaf node, we will encounter two possible situations:

1. The node is inside the search range – node A in Figure 12.18 is an example of this. Everything to the right of this node must be inside the range. We know this because everything to the right has a greater Morton code than A, but a smaller one than S. Therefore, we can trace down this right-hand branch and add all the leaf nodes we encounter to our results.

2. The node is outside the search range – node B in Figure 12.18. We can ignore the right-hand child of B because everything down this branch will have a larger Morton code than B and so will fall outside the search. We therefore ignore this and look at the left child node.

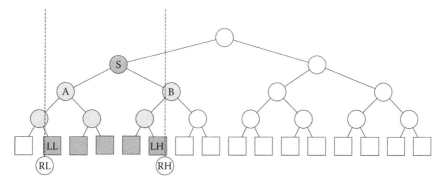

FIGURE 12.18
Principle of a range search on a balanced binary tree.

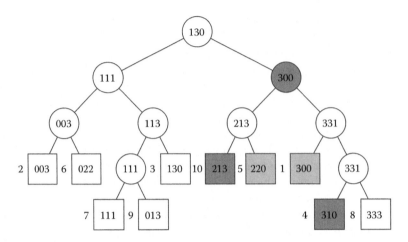

FIGURE 12.19
Range search using a balanced binary tree.

When we encounter a leaf node, the search is complete. The same procedure applied to our tree will produce the result shown in Figure 12.19.

Let us see how this procedure translates into code. From the root node, we need to move down the tree looking for the first node which falls inside the range. This can be done as follows:

```
Procedure find_split(node)
while node is NOT a leaf and MORTON[node] OUTSIDE RANGE
  if MORTON[split] < SL
    find_split(RIGHT[split])
  else
    find_split(LEFT[split])
return(node)
```

The procedure checks the node it is supplied with to see whether its Morton value (Morton[node]) is outside the search range. If it is, then depending on whether it is too low or too high, it calls itself again beginning with the appropriate child of the current node. When the test on the Morton code finds that it is not outside the range, we have found the node we want and the procedure returns the split node to the calling program.

Once we have found the split node, the procedure to search the left branch from this node is

```
Procedure search_left(node)
if node is a LEAF node
  report node
  return()
else
  If MORTON[node] > SL
```

```
        report_subtree(right(node))
        search_left(left(node))
else
        search_left(right(node))
```

This procedure will be passed a pointer to the left child of the split node. If this is a leaf node, then the procedure reports this and finishes. If it is a non-leaf node, then it looks to see whether its Morton address is inside the range or not. If it is, then it calls a procedure (report_subtree), which simply lists all the leaf nodes it finds under the right child of the current node, and then it continues to traverse the tree starting at the left child of the current node. If the Morton address test finds that we are outside the search range, the procedure simply proceeds to look down the tree starting at the right child of the current node.

We have now seen how by using a 1D code for spatial data we can use the efficient tree-based algorithms for searching. However, the trees which we have looked at so far have been binary ones, in which each node can have only two children. This means that if the data being stored changes, the tree will need to be changed quite a lot to maintain its sorted structure. For example, if a new value of 0112 needed to be inserted into the tree in Figure 12.19, all the nodes below the current 0111 would have to be altered. There is a second problem caused by the fact that the number of levels in a binary tree can become very large. The number of levels is referred to as the depth of the tree and with binary trees is approximately $\log_2(N)$. This means that with just 1024 data points, the tree will have 10 levels. This is not a problem if the tree can be stored in memory, but if the database is large, it may have to be stored on disk. This means that each stage of an algorithm such as the range search which involves retrieving data from a different level of the tree may potentially mean accessing the disk, and as we have seen each individual disk access is relatively slow.

To address these issues, many databases use a variant of the binary tree, called a B-tree, which is structured as shown in Figure 12.20.

The first difference from a binary tree is that each node can hold more than one value. In Figure 12.20, the root node has two values which, as well as storing data, or pointers to data, act as indices to the child nodes in much the same way as a binary tree. The left child contains values which are all smaller than A, the middle child values between A and B and the right child values which are greater than B. Each child node in this case has 10 entries, but this is simply

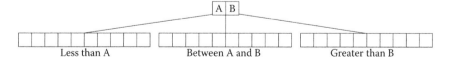

FIGURE 12.20
Structure of a B-tree.

to make it possible to produce a diagram. In practice, nodes in a B tree are designed to be large enough that they fill a significant portion of a disk block and may have of the order of 100 entries. What is more, each node can also have child nodes. For example, the entries in the left node of Figure 12.20 would have 10 values in numerical order, and these would act as indices to 11 child nodes, each with 10 entries. The root node in Figure 12.20 only has two entries, but it could also have 10 entries, and thus 11 children. This B-tree therefore has space to hold over 1000 data values in a tree which is only three levels deep.

The price which is paid with a B tree is that when a node is returned, you know that it will contain the value you are looking for, but you will need to do a linear search through the values in the node to find it. However, because disk access is so slow, searching N values in memory is much faster than doing the log(N) disk accesses it could take to find the same value in a binary tree which was stored on disk.

To understand how the B-tree might work to store our Morton codes, we will use a much smaller tree as shown in Figure 12.21. This is known as a 2–3 B-tree because each node has up to two values and can have up to three children.

To understand how a B-tree works, it is helpful to consider how the tree in Figure 12.21 was built. There are various ways to build a B-tree, one of which is to start with an empty tree and add the data values one at a time. Here, are the values in numerical order:

| 0003 | 0002 | 0111 | 0113 | 0130 | 0213 | 0220 | 0300 | 0331 | 0333 |

Each node can have two values and so the first node will contain the values 0003 and 0022. The question is what do we do when we add value 0111 (Figure 12.22a).

When a node becomes full, the middle value of those already in there and the new one is taken and used as a parent node. All those to the left of this middle value are put in the left child of the new parent, and all those to the right (including the new one) in the right child as shown in Figure 12.22b. To ensure that this split will always produce child nodes of equal size, the number of values allowed in a node is set to be an even number. If we define this number to be 2D, then when we try and add an extra value we will create

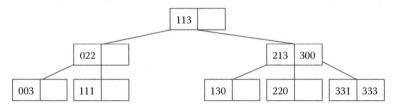

FIGURE 12.21
Storage of Morton codes for points in 2–3 B-tree.

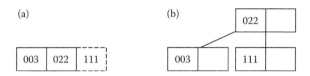

FIGURE 12.22
Splitting a node when it becomes full. (a) Node already contains 003 and 022 so there is no space for 111. (b) The node is split into two child nodes, one each for 003 and 111, and a new parent node at a higher level of the tree for 022.

two children of size D plus one new parent. What is more, the children will be half full, by definition. This means that the nodes in a B-tree will generally be at least half full, which makes the tree efficient in terms of the amount of data retrieved per disk access.

The next two values to be added to our tree are 0113 and 0130. 0113 can initially be placed in the same node as 0111 but 0130 will cause this node to become full and so it will be split. The middle value is 0113, but this time we do not need to create a new parent node because 0113 can go in the same node as 0022 giving the situation shown in Figure 12.23.

The addition of 0213 and 0220 will produce the situation shown in Figure 12.24.

The addition of 0220 means that the node containing 0130 and 0213 needs to be split and 0220 must be made into a parent node. However, it cannot be put with 0022 and 0113 because this node is already full (Figure 12.25).

The solution is that we simply apply the same rule as before – the node containing 0022, 0113 and 0213 is split, and the middle value – 0113 – becomes a new parent node, adding a level to the tree. 0130 and 0220 now become child nodes for 0213 as shown in Figure 12.26.

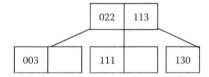

FIGURE 12.23
Another node split but this time the tree does not increase in depth.

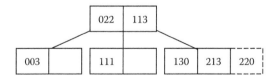

FIGURE 12.24
The addition of 220 means that two nodes will need to be split and another level will be added to the tree.

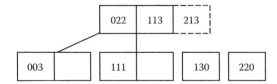

FIGURE 12.25
Stage 1 in the split – 213 is moved up and 130 and 220 are separated.

The remaining values can be added without the need of any further levels and we finish up with the complete tree (Figure 12.21).

Before we leave the topic of indexes, let us look at another of the many uses which have been found for space-filling curves. As well as being used as the values in an index, they can be used to help construct other indexes such as the R tree. Earlier in this chapter we identified that the way in which the objects were grouped together into the rectangles was important and that the ideal was to group neighbouring objects together. One of the parameters which must be set for an R-tree is the maximum number of objects per leaf node. This is usually chosen so that all the data for one leaf node will fit onto one disk block and can therefore be retrieved with one disk access. For the purposes of illustration, we will assume that three is the absolute limit in this case. One way of creating an R-tree is to sort the data by one coordinate and simply fill up the rectangles using the points in this sequence. If we do this for our data points, we get the result shown in Figure 12.27a.

Since there are 10 points, the last leaf only contains 1 point, which is rather inefficient, so we might modify our rule for filling the rectangle. For instance, we could decide each rectangle should have either two or three points. Once a rectangle has two points, we must decide whether the next point goes in the current rectangle or in a new one. The ideal is to group points which are as close as possible and this means that the size of the rectangles should be as small as possible. Therefore, in deciding which rectangle to assign our point to, we could consider the rectangle areas which would result in either case and pick the solution which gives the smallest. Such an approach will lead to a solution such as the one in Figure 12.27b.

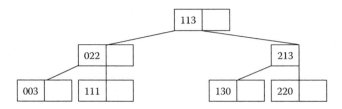

FIGURE 12.26
Split complete – 113 is the new root and 213 is the parent of 130 and 220.

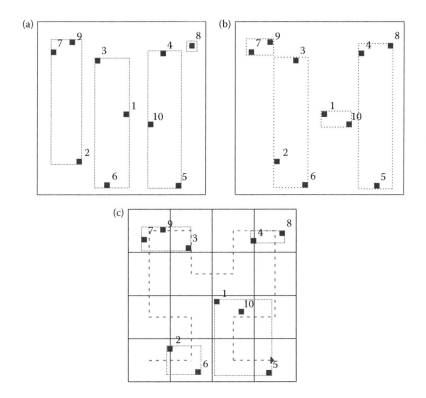

FIGURE 12.27
Three methods of assigning points to rectangles in the construction of an R-tree. (a) Using the
X coordinate of the points. (b) Using the X coordinate but minimising rectangle area. (c) Using
the Hilbert curve.

However, both these solutions suffer from the same problem we saw with
the row order example earlier. A range query which spans a large X range
is likely to intersect with all four rectangles, and therefore be relatively inefficient. We really need rectangles which are compact in both X and Y.

One way of doing this is to use a space-filling curve. Figure 12.27c shows
the points with a grid superimposed in which the grid cells are numbered in
Hilbert sequence. We now apply exactly the same approach as before, filling
each leaf with either two or three nodes resulting in a much better grouping
of the points.

12.6 Caching

So far, we have considered techniques of indexing and data organisation on
disk which speed up access to data. The final strategy which is used to speed

up data access is called caching, and describes a range of techniques for keeping a copy of data once it has been retrieved in case it is needed again. Caching is something that, depending on when you are reading this, you may be able to see in operation for yourself. Use a web browser to access Google Maps. Use a spatial search, but select somewhere that you would not normally look at such as another country or another part of your own country. As the first map draws on the screen, you should see it appear in rectangular blocks. If you pan sideways quickly, you should see the same effect; but, if you then pan back the way you have come, you probably would not get the blocky effect. Now, zoom in and notice how it takes a little while for the system to draw the image at the new level of detail. Zoom back out and you may notice the same thing, but when you zoom in it will draw a little more quickly.

What is happening here is that when you first enter your search, the system needs to get the appropriate data from the server, which means a transfer of information across the Internet. As we saw at the start of this chapter, this is the slowest form of data transfer and so having retrieved the data, the system stores it somewhere where access is quicker in case it is needed again. This is why when you return to the same view, this will load more quickly than it did at first. Just where the cache is located will depend on how the system is configured but if possible it will be stored on a disk drive on the computer you are using because this will provide the fastest response when the cache is accessed.

Caching has a large effect on the response time of the system and is not restricted to spatial data access. Most browsers will cache information from all your searches on the Internet. To see how much effect this has, see whether your browser allows you to clear the cache or to set it to a very small amount of space. Just where the appropriate option is located will depend on the browser and which version you are using; but, if you can clear out the cache and then go to a favourite site or your home page, you will notice how much more slowly it loads at first.

Caching is very important for disk access, but actually operates at all levels of the system. At the start of this chapter, a very simple hierarchy of Internet $\to$ Disk $\to$ Memory $\to$ CPU was presented. While essentially correct, this is much simpler than most computer systems in practice. For instance, a modern CPU will consist of a series of levels, each one a little faster, but able to store a little less data than the previous one, until the main processor and its registers are reached. Caching is used at all levels of the transfer of both instructions and code from memory to the processor.

One of the key issues with a cache is how to decide what should be retained there. This becomes an issue once the cache begins to fill up because eventually some more information will be retrieved. If this is to be saved in the cache, something else will have to be deleted to make room. The problem with deciding what to delete is that you need to know what data will be required in the future. For a programmer writing a program, this is a possibility and one of the issues to be considered when writing programs which access large datasets is to make sure that they use the cache efficiently.

A nice example of this can be given using one of the simplest algorithms we have considered in the book, namely, the calculation of area when the data are stored in a raster array. Below are two versions of the pseudocode which will implement this:

```
Program ROWMAJOR
Array RASTER[row][col]
For each row
   For each col
      If RASTER[row][col] = 'A' count = count + 1
Area = count*pixelsize

PROGRAM COLMAJOR
Array RASTER[row][col]
For each col
   For each row
      If RASTER[row][col] = 'A' count = count + 1
Area = count*pixelsize
```

The logic of the two programs is exactly the same – consider each array element in turn, see whether it has the appropriate value, and, if so, add 1 to the count of pixels. In efficiency terms, both programs will run in $O(n^2)$ time. The difference is that program ROWMAJOR takes a row at a time and processes all its columns, while COLMAJOR takes a column and processes all its rows.

However, both computer memory and disk storage are 1D and so in storing a 2D array, there are two possibilities:

- Store it a row at a time
- Store it a column at a time

With the 2 × 3 array shown in the below figure

1	2	3
4	5	6

the data values would be stored in the following sequence on disk and in memory:

Row major	1	2	3	4	5	6
Column major	1	4	2	5	3	6

This simple example makes it easy to see that if program ROWMAJOR is used with data stored in column major order; it will process the pixels in the order 1, 2, 3, 4, 5, 6 and that these are not stored in sequential memory locations. In fact, if that array size is large enough, they may not even be on the same page, which means the system would either have to load all the pages into memory at once, using up a large part of the cache, or swap pages in and out all the time, slowing down the processing.

However, a lot of caching is done by the operating system as a general mechanism to increase speed and in this situation, there is no way to control, or predict, what data will be required next. One solution to this is to perform some analysis on what the recent pattern of data usage has been and use this as a guide. For instance, if a page of data has been retrieved, but only accessed once, while another has been accessed several times, it would seem to make sense to retain the second one in the cache rather than the first one. The difficulty is knowing just what to measure and how to use this information to make the decision. Two of the common strategies which are used are known as 'least recently used' and 'most recently used' or LRU and MRU, respectively. As their names suggest, these both keep track of when each page in the cache is accessed, but take the opposite decision on what to discard. With an LRU strategy, the page which has stayed the longest in the cache without being used is discarded, on the basis that it does not appear to be needed anymore. With an MRU strategy, the most recently used page is discarded on the basis that programs often use information once and then move on to other things. Depending on the type of programs being run, either of these may turn out to perform well – or badly!

Alternatively, it can be argued that the cost of keeping access statistics itself takes time and storage and it is better to make the decision some other way. One option, which can be surprisingly effective, is simply to pick a page to discard at random. Another approach is to discard the largest item on the basis that it makes most room for new data.

One feature that all the techniques discussed in this chapter have in common is that there is no single 'best' solution to any of the problems being addressed. The success of indexing, storage and caching all depend on exactly what data are being processed and how. This means that the analysis of the success of different approaches tends to consider not whether the approach is guaranteed to be optimal because it is not possible to determine this, but how well it performs under various conditions ranging from the best to the worst. The approach which is selected may well be one which is rarely the best but which is never the worst and which can therefore be expected to perform reasonably for the majority of conditions. The final chapter of the book will consider a further set of problems for which it is often not feasible to produce optimal solutions but where it is possible to produce solutions which are good enough to be useful.

Further Reading

The difficulty of how best to store spatial data is nicely summarised in an early paper by Nagy (1980). Both Worboys and Duckham (2004) and Jones (1997) provide good descriptions of the topics covered in this chapter, including some techniques that are not described here, and van Oosterom (1999) provides a good review. Both versions of the NCGIA Core Curriculum have units on spatial indexes (http://www.ncgia.ucsb.edu/pubs/core.html). Cormen et al. (1990) is a good place to start for a description of various tree structures for non-spatial data, including binary search trees. De Berg et al. (1997) describe the use of k-d trees for range searching, and also methods for searching in non-rectangular windows. As with anything related to quadtrees, the two books by Samet (1990a,b) give a very thorough coverage. Samet (1990a) gives more details on finding which leaves in a quadtree are neighbours, for the more general case when the leaves are not all the same size. Samet (1990b) describes the point quadtree, which is similar to the k-d tree and its application to range searching. The use of space-filling curves for clustering and indexing data has received a lot of attention, not least because the concepts can be applied to any data which has more than one dimension, which can include non-spatial data which must be indexed by multiple attributes. The first part of Wang and Sun (2002) provides a very nice summary of the link between indexing, clustering and retrieval speed. Abel and Mark (1990) and Goodchild and Grandfield (1983) discuss the relative merits of a series of different scan orders for a range of queries and Bugnion et al. (1997) provide a general review of scan orders. More recent reviews, with a focus on database querying, are provided by Chen and Chang (2005), Lawder and King (2001) and Moon et al. (2001). Slightly more unusual is Dutton's (1999) work on the use of a triangular mesh for providing a uniform locational code for any point on the Earth's surface. There is also a web page (http://www.spatial-effects.com/SE-Home1.html) which illustrates these ideas (along with a variety of other things spatial). Comer (1979) provides a good overview of the B-tree and some of its variants and the site (http://slady.net/java/bt/view.php) provides a nice animation of how the tree is modified as elements are added and deleted. Li et al. (2012) discuss how an analysis of the way in which people use online mapping services may be used to improve caching behaviour.

13

Heuristics for Spatial Data

In Chapter 12, we saw how finding the shortest route could be speeded up by incorporating a heuristic estimate of the distance from each point to the destination. The term 'heuristic' can be applied more generally to any approach which does not guarantee to find an optimal solution in the way that an algorithm does, but that in most cases will find a good solution and may often do this more quickly than the optimum algorithm. In this chapter, we are going to look at some problems for which heuristics provide the only solution because there are no known algorithms which will solve the problem in a reasonable time.

To illustrate the general nature of the problem, let us consider the first of our two examples, the travelling salesman problem (TSP). Given a set of points in space, the TSP aims to identify the shortest possible route which visits each point exactly once and returns to the start. As with any problem, one approach is the brute force algorithm – look at all possible routes which visit each place once and find the shortest. The number of possible solutions for the TSP is easy to calculate. With n points, from the first point, there is a choice of $n - 1$ points to travel to, and from there $n - 2$ and so on. This gives $(n - 1).(n - 2).(n - 3)\ldots1$, which is written as $(n - 1)!$. In fact, if the cost of travelling from B to A is the same as travelling from A to B, this number is halved because a route which goes A–B–C–D is exactly the same as one which goes D–C–B–A. So, the number of solutions is given by

$$\frac{(n - 1)!}{2} \tag{13.1}$$

which clearly means that the TSP is $O(n!)$. A characteristic of the $n!$ function is that it shows exponential growth, which means that as n increases, not only does the size of the problem increase but also the rate at which this size increases also increases! This can be seen from the table of the first few values of $n!$ in Figure 13.1.

At each stage, $n!$ increases by a factor of n, so that 4! is four times larger than 3! Because of this, $n!$ has reached three million with a problem size of only 10.

This growth makes the brute force approach to solving such problems impractical for all but the smallest cases. If we have 20 cities, then the number of possible routes is approximately 6e16 (i.e., 6 with 16 zeroes; the e stands for exponent and is a commonly used shorthand for writing floating point

n	$n!$
1	1
2	2
3	6
4	24
5	120
6	720
7	5040
8	40320
9	362880
10	3628800

FIGURE 13.1
Value of $n!$ for n from 1 to 10.

numbers). With a computer which can calculate the length of a route in a millisecond, it will take around two million years for the brute force method to test all the possible routes and find the best. This means that the brute force algorithm is out of the question in most cases. The two problems we will consider in this chapter both have this $O(n!)$ characteristic and in both cases, no algorithm has been discovered which will provide an optimal solution in a reasonable time. We will return to the question of whether such an algorithm might ever be found later in the chapter. For the moment, let us look at how heuristics can be used to tackle such problems.

13.1 Travelling Salesman Problem

One of the intriguing things about the TSP is that there is a remarkably similar problem, for which algorithms do exist. This is called the minimum spanning tree (MST) problem, and the only difference is that the route only needs to connect all the points using the shortest path – it does not need to return to the start. More formally stated, the MST problem is to find the shortest path which allows someone to travel from any point to any other point. To see why this is different from the TSP, consider Figure 13.2. The route on the left is the TSP solution for these four points. However, to solve the MST problem, the direct link between 1 and 4 is not needed since to get from 1 to 4 it is possible to go via points 2 and 3. The route on the right is the MST for these four points – all the points are connected and the total length of the path is as small as possible.

It might seem that since the difference between the problems is so slight, the MST should form a good solution to the TSP. It is true that it often forms

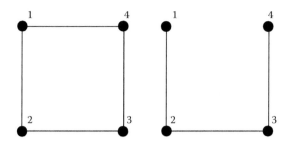

FIGURE 13.2
Solution to the travelling salesman problem (left) and minimum spanning tree (right) for four points.

a good approximation, but it is not hard to show that it does not form a solution to the problem. Figure 13.3 shows four points, and on the left the MST, which is 15 units long. The only way to convert this to a closed route is to join the first and last points, as shown in Figure 13.3b giving a total distance of 25.3 units. However, the TSP route for these points, which is shown in Figure 13.3c, and is only 22 units in length.

To explore some of the approaches which might be taken to producing solutions to the TSP, we will use the points shown in Figure 13.4, which we can imagine as a series of cities in an unusually regular layout. The distances between cities is 3 units in the N-S direction and 4 units in the W-E direction, distances which have been deliberately chosen to make the diagonal distances 5 units and thus produce nice, integer route lengths. No roads have been shown in the diagram, and it is assumed that it is possible to travel between any pair of adjacent points.

Figure 13.5 shows some possible solutions to the TSP for these points. The best routes (Figure 13.5a and b) are those which make the greatest use

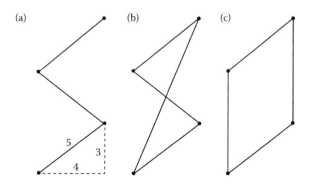

FIGURE 13.3
Illustration of why the minimum spanning tree does not form an optimal solution to the travelling salesman problem.

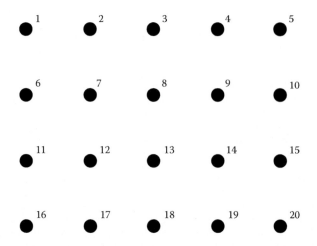

FIGURE 13.4
Street network for travelling salesman problem. It is assumed that roads exist between all neighbouring points both orthogonally and diagonally.

of the shorter N-S roads and the worst (Figure 13.5c) is the one which uses the long diagonal roads most.

An obvious heuristic for finding a solution to the TSP is to start at a point, and from there move to the closest, unvisited point. If we begin at point 1, this will take us down to point 16, along to 17, and then up to 2, along to 3 and so on. Eventually, we will reach point 20 and will hit a problem. What we have actually done is trace the MST and we have exactly the problem shown in Figure 13.2. The only way to close the tour is by joining the furthermost two points together and this gives a total distance of just over 83 for the route. As so often with spatial data, the simple approach does not produce a good solution – we are going to need to be a little more clever in designing a good heuristic.

The design of heuristics is rather similar to the design of optimal algorithms and some of the strategies used for designing good algorithms will lead to good heuristics. A strategy which is commonly used is to solve the problem

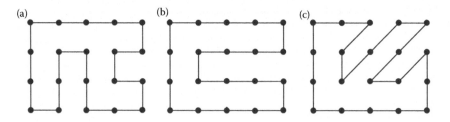

FIGURE 13.5
Some of the routes through the network that visit each point once. (a) Length = 70. (b) Length = 74. (c) Length = 78.

with a small subset of the data, and then add in the additional points. This was the format of the incremental sort algorithm (Chapter 10) and also of the algorithm for creating a Delaunay triangulation of a set of points (Chapter 9). In the case of the travelling salesman, one approach would be to take two of the points and find the shortest route between them. Then, we take another point at random, and find the shortest detour from the existing route, which will include the new point.

Assume that our first points are 1 and 7. There are two equally short routes, so we pick the one via point 2 at random. The next point is 19. This is easy and our route so far is shown in Figure 13.6. Note that we have had to pass through several other points on our way, so we add these to the current route.

The next two points are 14 and 16. These are close by the current route, so might not be expected to cause problems, but as we shall see they do. Consider point 14. One way to generate a detour is to replace one link in our current route with a route which visits the new point. The obvious link to replace is the one between the two closest points – 18 and 19. The shortest route between 18, 19 and 14 includes one of the diagonal links as shown in Figure 13.7. With point 16, the two closest points are 12 and 17 and replacing their link with a detour to 16 adds another diagonal to the route.

The sequence of points in this case was chosen to illustrate some of the problems with this heuristic. However, in general, any system which picks points at random is sensitive to the particular sequence of points.

An alternative approach is to try and consider what the characteristics of a good route would be. The ideal is a perfectly circular route, since this would contain no deviations at all. This is impracticable of course, but perhaps rather than starting with points picked at random we could start with the closest we can come to a circular route, which is what is called the convex

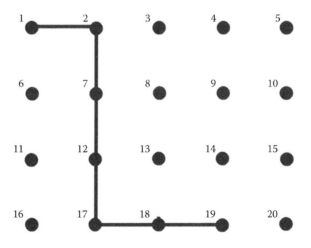

FIGURE 13.6
Incremental heuristic after 3 points.

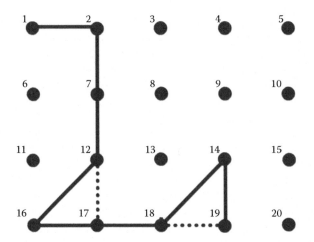

FIGURE 13.7
Incremental heuristic after 5 points.

hull (Figure 13.8). This term is used to describe the smallest polygon which encloses all the points. Now, rather than picking the remaining points at random, let us add the point which causes the least deviation from this route. In fact, all the remaining points can be added by creating a deviation from two of the points on the outer route, giving a final route as shown in Figure 13.9. The length of this route is 74, which is about halfway between the optimum and the worst routes in Figure 13.5.

The problem with the heuristics so far is that each additional point has been considered in isolation. For instance, if we had added points 7, 8, 12 and

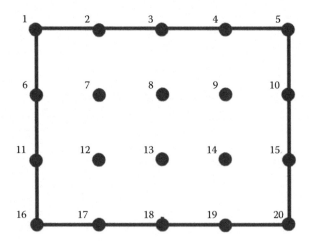

FIGURE 13.8
Incremental approach starting with convex hull.

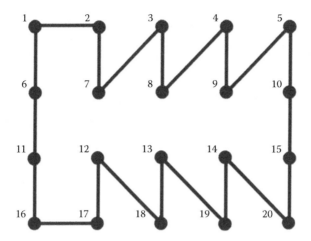

FIGURE 13.9
Route after adding all points to outer route shown in Figure 13.8.

13 at the same time as a diversion between points 2 and 3, this would have produced a shorter route still. However, it is much easier to produce heuristics which consider one point at a time than the ones which consider several points simultaneously. One of the strengths of algorithms, such as the incremental algorithm for creating a Delaunay triangulation is that it allows the situation around each individual point to be dealt with but in a way which guarantees an overall optimum solution. The reason that this is possible in this case is that the characteristics of the optimum solution are known. It is known that the Delaunay triangulation is an optimal triangulation of a set of points, that there is a unique Delaunay triangulation of any given set of points and what many of its properties will be. The same is not true of the TSP, which is part of the reason that there is no algorithm which is guaranteed to produce an optimal solution.

13.2 Location Allocation

Our second example will be the location allocation problem. Imagine we have the resources to locate two ambulance stations which need to serve a local area in which the population is spread between some large towns and smaller villages. Where should we locate the stations to best serve the population? This is one example of the location allocation problem in which the central problem is to identify the optimum location of sources to provide goods or services to customers. The problem comes in a variety of versions. For instance, it may be possible to locate the services anywhere or we may be limited to a few potential sites. Travel between services and customers may be in any direction or restricted to travel along a network. Each source may have a finite ability to provide services or may be regarded as having infinite resources.

However, even in its most limited form, the number of possible solutions to the location-allocation problem grows very quickly as the size of the problem grows. We will be considering the simplest form of the problem in which we have to select the best R locations for the sources from a number of possible candidate sites N. The expression for the number of ways of selecting R items from a list of N is given by Equation 13.2

$$C = \frac{N!}{R!(N - R)!} \tag{13.2}$$

In a moment we are going to consider a simple problem in which $N = 9$ and $R = 2$, in which case the number of combinations is 36. However, if N is 100 and R is 10, the number rises to 1.73e13. The number of combination grows in a combinatorial fashion if either N or R rises and the rate of increase also rises as the absolute values of N and R rise. For instance, increasing N by 1 means there are N more combinations because

$$(N + 1)! = N.N! \tag{13.3}$$

As with the TSP the brute force algorithm is clearly out of the question.

To consider one of the heuristics used, we will use a very simple version of the problem. Figure 13.10 shows nine locations equally spaced along the X axis, which means that in considering location, we only have one dimension to worry about. To simplify things even further, the nine locations all act as destinations and can all potentially be sources. Travel is only possible along the roads shown in Figure 13.10 and since we are dealing with a network, we will refer to the numbered locations as nodes from now on.

The problem in this case is to use two of these nine nodes as sources to supply goods to all nine nodes, where the cost of supplying the goods is determined by the distance between the supply and destination nodes. This is a very simple location-allocation problem, but this will allow us to introduce some important ideas about how more realistic problems are solved. As this is a small example, we can draw up a table of the cost of all the possible pairings. First, we would produce a table of the distance between all pairs of nodes as shown in Figure 13.11.

FIGURE 13.10
One-dimensional location allocation example. The nine points can act as both sources and destinations and are all 1 unit apart in the X direction.

	1	2	3	4	5	6	7	8	9
1	0	1	2	3	4	5	6	7	8
2	1	0	1	2	3	4	5	6	7
3	2	1	0	1	2	3	4	5	6
4	3	2	1	0	1	2	3	4	5
5	4	3	2	1	0	1	2	3	4
6	5	4	3	2	1	0	1	2	3
7	6	5	4	3	2	1	0	1	2
8	7	6	5	4	3	2	1	0	1
9	8	7	6	5	4	3	2	1	0

FIGURE 13.11
Distance between each pair of nodes in Figure 13.10.

From this, it is then possible to work out the total cost of each possible solution. For instance, if we select 1 and 2 as our sources, which we will represent as {1,2}, the cost is worked out as follows:

```
For each node n from 1 to 9
    Find distance from 1 to n in the table
    Find distance from 2 to n in the table
    Choose the shorter and add it to the total
```

The calculation in this case is

```
Cost {1,2} = 0 + 0 + 1 + 2 + 3 + 4 + 5 + 6 + 7 = 28
```

Doing this for all 36 possible pairings produces the table of costs shown in Figure 13.12. It is only necessary to fill in one half of the table since {1,2} is the same as {2,1}.

The table shows that three pairings share the lowest cost: {2,7},{3,7},{3,8}. Another way to visualise the results in Figure 13.12 is as a surface as shown in Figure 13.13.

Thinking of the costs under different combinations as a surface in this way is very common in tackling optimisation problems where it is referred to as the solution space for the problem. What the solution space does is show how the costs of the solution change as the parameters change, or in this case as the selection of nodes changes. It can be seen that the solution space has a clear low point, because this example has an optimum solution and that solutions which are close to this, such as {2,6}, have costs which are also low. One way to think about the location-allocation problem therefore is that what we are trying to do is find the minimum point in the solution space. The general approach taken by most heuristics is to start with an initial solution chosen at random, and then repeatedly change this with the aim of finding a lower cost solution.

	1	2	3	4	5	6	7	8	9
1									
2	28								
3	22	22							
4	17	17	18						
5	14	13	14	16					
6	12	11	11	13	16				
7	12	10	10	11	14	18			
8	13	11	10	11	13	17	22		
9	16	13	12	12	14	17	22	28	

FIGURE 13.12
Cost of all possible combinations of two points from the nine. Only the lower half of the table is shown because the cost for {2,1} would be the same as for {1,2}.

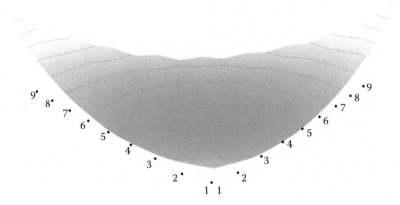

FIGURE 13.13
Results from Figure 13.12 depicted as a 2D surface.

To see how this works in practice, we will consider an algorithm for location-allocation problems first suggested by Cooper (1972), which he called the alternating transportation-location heuristic or ATL. It works as follows:

1. Allocate two points as sources at random.
2. Assign each destination to its nearest source point. This produces a table of which destinations will be supplied by each source.
3. *Repeat*
   a. For each group of destinations, find the source which could supply them at the lowest cost. This may change the source points.

  b. Assign each destination to its nearest source. This may change the source which supplies each destination.

4. *Until* there is no change in the source/destination table.

The algorithm is most easily understood by seeing how it would work with our example and this is shown in Figure 13.14.

The algorithm first picks two nodes at random, and in this case, it has selected 1 and 2 (Figure 13.14, Initialize). Each destination is then allocated to its nearest source. This results in node 1 supplying itself and node 2 supplying all the other nodes.

On the first iteration (Figure 13.14, 1a), we first consider all the nodes currently being supplied by each source, and see whether having them supplied by a different source would reduce the total cost. Only one node is being supplied by source 1 and we cannot reduce the cost since it is zero. However, eight nodes are being supplied by node 2, and by supplying them by a node in the middle of the group we can reduce the cost. We therefore select node 5 instead of 2 to supply these nodes, which brings the total cost down from 28 to 16. Now that we have made a change to the nodes which are selected as sources, the second part of this iteration (Figure 13.14, 1b) is to reallocate all the nodes to their nearest node. Node 2 is now nearer to source 1 and so

Iteration	Sources	Destinations									Total Cost
		1	2	3	4	5	6	7	8	9	
Initialize	1	1									
	2		1	1	1	1	1	1	1	1	
Cost per node		0	0	1	2	3	4	5	6	7	28
1a	1	1									
	5		1	1	1	1	1	1	1	1	
Cost per node		0	3	2	1	0	1	2	3	4	16
1b	1	1	1								
	5			1	1	1	1	1	1	1	
Cost per node		0	1	2	1	0	1	2	3	4	14
2a	2	1	1								
	6			1	1	1	1	1	1		
Cost per node		0	1	3	2	1	0	1	2	3	13
2b	2	1	1	1							
	6				1	1	1	1	1	1	
Cost per node		1	0	1	2	1	0	1	2	3	11

FIGURE 13.14
Steps in the ATL algorithm.

is switched to be supplied by this node, which means that at the end of iteration 1, the costs have been reduced to 14. In the second iteration (Figure 13.14, 2a and 2b), this process results in the sources being changed from 1 and 5 to 2 and 6 and some reallocation of nodes to nodes, and the cost is further reduced to 11. The next iteration would not be able to improve on this and so the heuristic would terminate after that. The final solution of {2,6} is good, but not quite optimal. The problem is that to reach the optimal solution, node 4 would have to be allocated to 2, in which case the best supplier for the remaining nodes of 5–9 becomes 7, which leads to the optimal solution. However, since 4 is the same distance from 2 as it is from 6, the heuristic has no reason to make this switch. However, the point of heuristics is that they should find good solutions, not the absolute best solution.

The ATL heuristic always looks for the change that will produce the greatest improvement in the solution. This 'greedy' approach as it is known is equivalent to always moving downhill as quickly as possible in the solution space (Figure 13.13). This seems sensible, given that we are trying to find the lowest point in the solution space. However, not all solution spaces are as simple as the one shown in Figure 13.13 and more complex spaces can create problems for such greedy heuristics. Figure 13.15 shows a slightly modified version of the problem in which three changes have been introduced:

1. The links between the nodes now have different costs so that it takes twice as long to get from 3 to 4 as it does to get from 3 to 2.
2. There is an extra road which links 1 and 9.
3. This road has different costs depending on the direction of travel.

This complicates the problem because the additional road between nodes 1 and 9 means there are two possible routes between each pair of nodes. The solution space for the modified problem is shown in Figure 13.16, both in a table and graphically.

The solution space now has two low points, each containing two pairs of nodes as follows:

{1,5},{1,6} Cost = 14
{4,9},(5,9) Cost = 11

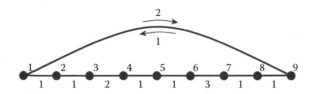

FIGURE 13.15
Modified network for location-allocation example.

(a)

	1	2	3	4	5	6	7	8	9
1									
2	22								
3	19	22							
4	15	17	20						
5	14	16	18	24					
6	14	15	17	22	25				
7	17	15	14	14	15	17			
8	17	15	14	12	12	14	22		
9	20	17	14	11	11	12	19	20	

(b)

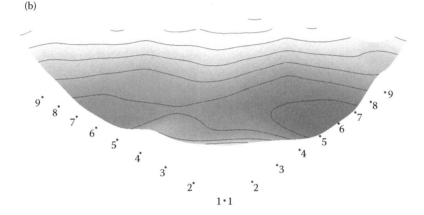

FIGURE 13.16
Solution space for the allocation of two supply nodes for the network shown in Figure 13.15. (a) Costs of solutions. The two minima are enclosed in boxes. A greedy heuristic which starts in a shaded cell will get stuck in the higher of the two minima. (b) Solution space.

If we use a greedy heuristic, this means that with some starting points, the heuristic will move downslope to the 'wrong' minimum. In fact, any of the starting points which are shaded in Figure 13.16a will result in the heuristic reporting the local minimum with a value of 14 instead of the one with a value of 11.

13.3 Metaheuristics

So far we have considered heuristics which are specifically tailored to the particular problem. However, there are also a number of what are called metaheuristics, which are methods that can be applied to a range of problems. TSP and location-allocation problem are just two examples of what are called optimisation problems in which the goal is to find the best solution to a problem which is too complicated to be solved directly. Such problems do not have to be spatial. For example, in economics, the problem of how best to spend your money to the best effect is a type of optimisation problem in which you have to weigh up the relative costs and values of a range of goods and services. What is interesting from a GIS perspective is that whatever the problem, finding an optimum solution is usually conceptualised in spatial terms, as trying to find the global minimum (or in some cases maximum) in a solution space. Underlying this is the idea that if the parameters are only changed slightly, the cost of the solution will also only change a small amount. In other words, the solution space is thought of as being smooth.

The solution space in the location-allocation problem had only two dimensions and so could be directly visualised. If the problem had been to choose four of the nodes to serve as sources, this would not have been possible because we would have had a four-dimensional (4D) space. Most real optimisation problems have solution spaces with far more dimensions than this. However, this makes no difference to the mathematics, which can be used to analyse a 4D surface as easily as a 2D surface. In terms of comprehension though, it is useful to think of a simplified version of the solution space as a surface with a series of hollows (i.e., local minimum points) and one hollow which is deeper than all the others and that is the one we would ideally like to find.

The real difficulty in finding the optimum point in the solution space is that we do not know what the solution surface looks like because we cannot calculate all the possible solutions. Given a starting solution, we can certainly proceed downhill until we can go no lower, but we have no way of knowing whether we have found the absolute minimum of the surface or have simply ended up in one of the other hollows. All we can really do is try and develop methods that search as much of the solution space as possible before deciding on what seems like an optimum solution. There are a number of methods which do this and we will consider simulated annealing (SA) as an example. Like a lot of metaheuristics, it is based on the principle that in order not to fall into the nearest low point from the starting solution you must initially accept some solutions which appear to be worse than the current one in the hope that this increases the chance of finding a better final solution.

SA is based on an analogy with cooling liquid metal to make it solid, a process which is called annealing. If the cooling is done slowly, the metal tends to form into crystals which will interlock and create a strong metal. If the cooling is too rapid, there is not enough time for the crystals to form and the resulting metal is therefore less strong. The analogy with optimisation is that in the early stages, the heuristic will allow changes which produce an increased cost, but that as time goes on the probability of doing this reduces until it eventually falls to zero. From there on, only changes which improve the cost are accepted.

The SA algorithm can be described as follows – each of the stages will be explained in detail afterwards:

1. Generate an initial solution.
2. Choose a neighbouring solution in the solution space.
3. Accept this as a new solution if either.
   - It reduces the cost.
   - It increases the cost but the 'temperature' of the system allows the swap.
4. Reduce the temperature.
5. Repeat 2–4 until the system has 'cooled'.

The first step is to generate a 'solution' to the problem at random. In the case of the TSP, this might mean simply generating a circular route which visits all the nodes. In the case of location-allocation, it might mean assigning sources to destinations at random. From here, a small change is made to the solution such as swapping two nodes in the current solution. In the case of the TSP, it is usually taken to mean that we consider swapping one of the nodes on the current tour with one of its immediate neighbours. With the simple location-allocation problem, it would mean swapping a node with one of the neighbours on either side. In a full 2D example, it would probably be taken as the 2D equivalent, which is one of the neighbouring points in the Delauney triangulation of all the points. As well as choosing spatial locations, many problems also have other parameters which need to be optimised such as the relative quantities to be supplied by each node in a location-allocation problem. In this case, a neighbouring solution might be one which keeps the same set of nodes, but makes a small adjustment to the relative demands on each node.

So, we have the idea that at each iteration, there will be a current solution with a cost C1 and a candidate new solution with a cost C2, which might be smaller or larger. Now, let us consider in more detail how steps 3 and 4 are implemented. The decision on whether to accept the new solution is based on the following relationship:

$$e^{\frac{-(C2-C1)}{T}} > R[0,1] \qquad\qquad (13.4)$$

where T is the temperature of the system and $R[0,1]$ is a random number between 0 and 1.

To explain what is going on here, let us first simplify this to the following:

$$e^{-k} > R[0,1] \qquad\qquad (13.5)$$

where

$$k = \frac{(C2 - C1)}{T} \tag{13.6}$$

Factor k depends on the change in costs $C2 - C1$ (ignoring T for a moment) and will be positive if the move would increase the costs and negative if it decreases the cost. Figure 13.17 shows the function e^{-k} for values of k from -5 to $+5$.

When k is negative, which means that the move would decrease the cost, the function is greater than 1. This means that no matter what random number in the range [0,1] is selected e^{-k} will be greater and the swap will be accepted. When k is positive, the function lies between 0 and 1; the smaller the value of k the higher the value of the function. Whether the swap is accepted now depends on whether the value of the function is greater than the random number. Clearly, the nearer the function is to 1, the more likely this is to be true. In summary, then Equation 13.5 means that

- If the swap decreases the cost, it will always be accepted.
- A swap may be accepted if it increases the cost, but this is more likely if the increase is small.

With this understanding we can return to the full form of the expression in Equation 13.4.

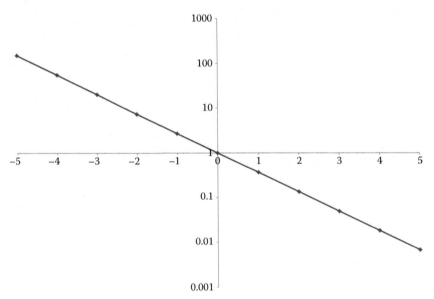

FIGURE 13.17
The function e^{-k} for values of k from -5 to $+5$. Note that the Y axis uses a logarithmic scale.

It is clear that as well as the difference in cost of the two solutions, k also depends on a second parameter T, which is regarded as the temperature of the system. When T is large, k will be small. In fact, if T is much larger than either $C1$ or $C2$, k will be small regardless of the value of $(C1 - C2)$, which means that almost any change to the solution is likely to be accepted. As T becomes smaller, k increases in size, so that e^{-k} moves towards zero and eventually the chance of a swap which makes the solution worse being accepted becomes practically nil.

It is now possible to write out the heuristic for SA in a more detailed fashion:

```
Set T to an initial value
Set a rate at which T is reduced: t
Generate an initial solution. This has cost C1
n = 1
Loop until n > nmax or C1 < GOOD_ENOUGH
      Select a neighbouring solution in the solution space.
      This has cost C2
      D = C2 - C1
      If exp(-D/T) > random(0,1) then
         C1 = C2
         n = n + 1
         T = T - t
```

In this code, `random(0,1)` is a procedure which returns a random number between 0 and 1. Id $C2$ has lower cost than $C1$, D will be negative, the exponential will give a value above 1, and the swap will be accepted. If $C2$ has a higher cost, the exponential function will produce a value between 0 and 1 and so it is a matter of chance whether the swap will be accepted. As T becomes smaller, the exponential result also gets smaller and so the chance of the swap being accepted falls.

Before we run this heuristic, there will be a number of decisions which have to be made:

- What should the initial temperature be?
- How quickly should it reduce?
- When should we terminate? After a fixed number of iterations? After a sequence of moves with no change? When the cost of the solution reaches an acceptable value?

There is no best answer to any of these questions. The only two strategies which we can use are to try it and see or look at what other people have done and use their experience to guide us. All this may seem a little ad hoc, but it is important to remember that we are dealing with problems where we have no other options, so this kind of practical solution is all that is available. And in practice, heuristics do provide good solutions to difficult problems.

They also reflect the way that people make decisions in everyday life. When we plan a route to the shops, we are happy with one that seems reasonably short – we do not spend hours with maps comparing endless options to shave a few metres off. We settle for a solution which is good enough for our purpose and this is what heuristics also provide in the case of GIS.

13.4 Computability and Decidability

Let us now return to the statement made at the start of the chapter that for problems such as the TSP and location-allocation, there are no known algorithms which can solve them in a reasonable time. The question of whether such algorithms can ever be produced is one of the great unanswered questions of computational analysis. In one sense of course, an algorithm does exist to solve these problems – the brute force algorithm which lists all the possible solutions and picks the best. However, this is impracticable for any but the smallest networks. The question is therefore not whether an algorithm is possible, but whether an algorithm which has a reasonable degree of complexity is possible. 'Reasonable' is normally taken to mean polynomial complexity, that is $O(n^2)$ or $O(n^3)$. All algorithms which have polynomial complexity or better are called class P algorithms. Any algorithm which can be run in polynomial time can also be checked in polynomial time. For instance, given the output of a sort routine, it is possible to check that the algorithm has indeed sorted the numbers in $O(n)$ time. Each number is checked in turn; if one is found to be out of sequence, the check fails.

Interestingly, although it may not be possible to solve a problem in polynomial time, it is often possible to check a proposed solution in polynomial time. Such algorithms are called NP, which stands for non-deterministic polynomial. Non-deterministic means that a solution to the problem can be found (the problem is not completely intractable), but that this solution may involve making an educated guess and then seeing if that has in fact worked. Polynomial means that this check can be done in polynomial time. The TSP and location-allocation problems are both examples of an NP problem – heuristics exist which will generate solutions, and it is possible to test whether these are in fact optimal.

It is known that all problems which belong to the class P are also NP. However, it is not known whether NP problems also belong to the class P. Why does this matter? If it could be proved that NP problems cannot be solved in polynomial time, then there would be no need to carry on trying to develop algorithms for their solution – we would know that heuristic approaches are the best we are ever going to get. Conversely, if it could be proved that NP problems did belong to class P, this would make it worth trying to find an algorithm for such problems.

But, is it possible to prove that a problem like the TSP can be solved in polynomial time without actually doing it? Surprising as it may seem, the answer is yes. The reason is that all computing problems, no matter how simple or complex, can be analysed and compared in a way which does not depend on their exact details. In Chapter 1, we introduced the concept of a Turing machine, which was an idea first proposed by English mathematician Alan Turing. The Turing machine is an imaginary 'computer', which consists of an infinite length of tape that can be read by a small, movable head. The tape contains sequences of symbols, and depending on the symbol which is read, and the 'state' in which the reading head is in at the time, the head will perform one of a series of actions – move left, move right, write a new symbol. The remarkable thing is that it has been shown that any form of computation, from adding two numbers to modelling the earth's climate, could in theory be programmed on a Turing machine. If this is the case, then all these problems must be in some way comparable with each other. If there were a set of computations which could not be run on a Turing machine, then these would form a completely different class from those that could. (As an interesting aside, many people believe that the human brain is capable of carrying out operations which cannot be programmed on a Turing machine, and that this means that current computers will never be programmed to reproduce human intelligence.) As a result of this comparability between computer programs, it is possible to analyse a particular algorithm and assign it to a class such as P and NP. It is also possible to show that certain problems, such as the TSP, can be converted to be the same as a different problem. To take a trivial example, it is easy to see that if we want to find out whether a point falls inside a rectangle, we do not necessarily need a special algorithm – we can use one which will determine whether a point falls within an area of any arbitrary shape. This will be less efficient of course, but that is not the issue here. Once we have an algorithm for problem A, we can use this to prove that an algorithm must exist for any other problem B, which can be converted to problem A. This conversion will not necessarily be a good solution in itself, but the proof that an algorithm must exist, means it is worth the effort searching for an efficient solution to problem B.

Let us return to the TSP. This belongs to a special set of NP problems called NP-complete. These are NP problems which can be shown to all be equivalent to one another. If it can be shown that an efficient algorithm exists to solve one of them, this would prove that efficient algorithms exist for all of them. Conversely, if it could be proved that no efficient algorithm exists for one of them, this proves that none exists for any of them. Proving which of these is true remains one of the great unsolved problems of computational complexity. Until it is solved, programmers will continue to work on better heuristics to provide practical solutions to problems such as the TSP.

Throughout this book, repeated reference has been made to results which have been proved by computer scientists, such as the fact that the Dijkstra algorithm is guaranteed to find the shortest path between two points. To give some sense of how such things are proved, we will end with a classic proof from Alan Turing. Turing was involved in the early development of electronic computers as part of the code-breaking efforts at Bletchley Park during the Second World War. The machines were designed to do a very practical job of computation but as a mathematician, Turing was also interested in the philosophical problem of what could and could not be computed. His interest in the difference between computation and intelligence led him to propose what is now known as the Turing test of machine intelligence. He proposed a test in which an operator was able to communicate with two subjects in a way which left them completely anonymous. One of the subjects was a computer and the other was a person. The operator is able to ask the subjects any question and if, from the answers, it is not possible to distinguish between the computer and the human, then the computer is deemed to be displaying intelligence. This has stimulated a lot of work in artificial intelligence, but to this day no computer has passed the Turing test. Incidentally, the Turing test is what lies behind the Captcha tools on websites that ask you to prove that you are a human being by copying out sequences of letters and numbers that have been made difficult to read by automated means.

One of the theoretical problems that Turing considered was whether it was possible to decide whether a computer algorithm was going to terminate or not. For example, look at the two programs in Figure 13.18.

They all have a similar structure in that variable n is first set to 1 and then a loop is set up which terminates when n has reached 10. Once the loop has terminated, the program stops. Program ONE will go through the repeat loop 10 times and then stop. Program TWO will never stop because n is never changed. Turing wondered whether it would be possible to write a program which could analyse the structure of programs such as ONE and TWO and automatically decide whether they would terminate or not. This may seem too difficult a problem to solve but in fact Turing was able to prove that such a program is impossible. He did this using a technique called 'proof by contradiction' in which you assume that what you want to prove is not true and

```
program ONE
n = 1
repeat until n > 10
  n = n + 1
stop

program TWO
n = 1
repeat until n > 10
  n = 1
stop
```

FIGURE 13.18
Example programs for the halting problem.

then see whether this leads to a logical contradiction. If it does, then you have shown the original assumption to be false. A simple and famous example is Euclid's proof that there is an infinite number of primes. A description of this is provided in a separate box for anyone who is not familiar with this kind of mathematical logic. Turing used the proof by contradiction by assuming that a halting program could be written and then considered what would happen if such a program were applied to itself. So, first, we assume that we have a procedure called terminate which takes the text of any computer program as its input. It returns the result TRUE if this second program will terminate and FALSE if it would not. In pseudocode

```
boolean terminate(program)
If (clever_test_shows_program_terminates)
return TRUE
Else
return FALSE
```

Just how the program makes the test has been left vague because in a moment we are going to show that terminate cannot exist. Assuming for the moment that the terminate program does work, we can then write a program which uses it to test the code for programs ONE and TWO above, that is

```
Program test
Print 'Result for ONE: ',terminate(ONE)
Print 'Result for TWO: ',terminate(TWO)
```

And when run, this would produce the output

```
Result for ONE: TRUE
Result for TWO: FALSE
```

PROOF BY CONTRADICTION

Euclid's proof that there is an infinite number of primes is a famous example of proof by contradiction. You start by assuming the opposite of what you assume to be the truth and then whether this leads to a contradiction or logical impossibility. As long as all your reasoning is correct, the only thing that can be incorrect is the initial assumption. Euclid's proof depends on the fact that prime numbers can only be divided by themselves and 1. For instance, number 3 cannot be divided by 2, which is the only other integer apart from 1. Dividing 3 by anything larger than itself produces a fractional result. Hence, 3 is a prime number. It follows from the definition that any number which is not a prime number must be divisible by a prime number. How do we know this? Imagine we have a number N which is not a prime number and we go through all the numbers smaller than N to see which ones divide into it. Assume none of these is a prime number. This now gives us a new series of numbers which have to be divisible by something because they are not primes. So, we repeat the process with all these. We cannot carry this on forever because we do not have an infinite number of integers between N and 1, so eventually we must find a prime number. Here is our first example of how mathematical logic can be used to prove propositions. Now, we can present the proof. However, before presenting it as a general case, let us consider it using some actual numbers.

We start by assuming that there are just two primes – 2 and 3. We create a new number

$$2 \times 3 + 1 = 7$$

We know that any number which is not prime must be divisible by a prime number. However, 7 cannot be divided by either 2 or 3 – it always leaves a remainder of 1. So, there are only two possibilities. Either there is another prime number that we have missed – or seven itself is prime, which is the case here. To show that this argument follows no matter how many primes we think we have found, we repeat the operation using the first six primes:

$$(2 \times 3 \times 5 \times 7 \times 11 \times 13) = 30031$$

In this case 30031 is not a prime but it is only divisible by two primes which are not in our current list – 59 and 509.

Now, we can generalise. We start by assuming that there is a fixed number of primes n. If the first is $p1$, the second $p2$ and so on, then you create a new number P as follows:

$$P = (p1 \times p2 \times \cdots pn) + 1$$

Remember that a prime number is one which is only divisible by itself and 1. We have assumed that we know all the prime numbers and if this is true, then P must be divisible by one of the existing prime numbers. However, if we divide P by $p1$, we will get a remainder of 1 and this will be true for all the other known prime numbers. This means that one of two things must be true. First, P could also be a prime number. Second, there could be another prime number which was not in the original list that divides into P. Whichever is the case, we have shown that there is another prime number and that our original assumption that we knew all the primes was false. This in turn is a proof that there is an infinite number of primes.

However, we can write another version of test that also calls terminate but which contains a clever trap as follows:

```
Program test(program)
If terminate(program) then
Loop forever
Else
Print TRUE
```

If we use test on ONE and TWO, we simply get the wrong result. For instance, if we use test(ONE), then terminate finds that ONE finishes, but then test goes into an infinite loop. If we use test(TWO), it finds that TWO would not terminate, but wrongly returns the answer that it will. These are essentially just bugs in test, which means it does not work properly. However, we can run test on itself – test(test). In other words, we can run test and supply as input the above code. Test is just a program like any other and so if terminate is working properly, it should be able to tell whether it will ever end. Program test starts by testing the code it is supplied with to see whether it will terminate. If it finds it will, it goes into an infinite loop. However, this means that test does not terminate and so the terminate procedure should return FALSE. However, if it does this, the test will terminate and so terminate should return TRUE.

We have reached a logical contradiction. No matter how we try and write the terminate procedure, it cannot produce the correct result in this case. This means that it is not possible to write a version of terminate, which is guaranteed to work because we have identified a situation where it cannot ever work. In formal terms, this means that we have proved that the halting problem is not computable.

This may all seem a long way from GIS and in one sense it is. There are several other problems which, like the halting problem, have been shown to be non-computable, but they tend not to fall within the realm of practical data processing. However, the halting problem nicely demonstrates how the techniques of mathematical logic can be applied to the analysis of algorithms. It is these techniques which allow computer scientists to determine whether an algorithm is optimal and also to show that NP-complete problems are all equivalent. This means that some very important problems in the field of GIS, such as the TSP and location-allocation, fall within the realm of an important area of study within computer science. Since there are currently no known algorithms for solving these problems, it also means that a good deal of work is undertaken to try and develop heuristics which can provide practical solutions to them.

Further Reading

David Harel's book *Computers Ltd. What They Really Can't Do* (Harel 2000) is an excellent and very readable account of the issues of what is and is not computable. It also covers many of the other issues which have been covered in this book, such as big-O notation and the design of algorithms. Georgia Tech has an excellent website devoted to the TSP (http://www.tsp.gatech. edu/index.html) which provides a history of approaches to the problem, key references and examples of notable solutions. The location allocation problem is covered in one of the units in the second version of the NCGIA core curriculum (http://www.geog.ubc.ca/courses/klink/gis.notes/ncgia/ u58.html). Cooper (1972) first described the ATL heuristic for the location-allocation problem, which he called the transportation-location problem. Liu et al. (1994) describe the application of SA to this problem, while a more recent paper by Arostegui et al. (2006) reviews a range of heuristic methods. Blum and Roli (2003) provide a good summary of metaheuristics overall. The metaheuristics project has a web page which describes the work undertaken under an EU grant between 2000 and 2004 on a range of issues related to metaheuristics. One of the problems described, the vehicle routing problem, is essentially a combination of the TSP and location-allocation since customers must be served from a set of fixed nodes but also using an efficient route. Alan Turing's name has occurred a number of times in this book. This website, which is maintained by Andrew Hodges, gives not only more details of this remarkable man but also a lot of background on the areas in which he made significant contributions, including the theory of computation and the mathematical question of decidability, which inspired Turing's work on computability.

Conclusion

A few themes have emerged in the course of the book and it is worth finishing by trying to pull these together. The first is the set of problems caused by the 2D nature of spatial data. This is not a problem which is unique to GIS. Computer graphics and computer-aided design (CAD) software also have to handle 2D data, and in fact uses many of the same data structures and algorithms as used in GIS. Where GIS is perhaps different is in the need to handle large volumes of data and to be able to undertake what are effectively database operations. Where specialist software packages are entirely suitable for computer graphics and CAD they present some problems for GIS. For example, specialist GIS software cannot take advantage of the well-tested and very efficient facilities of modern database software. This approach also makes it more difficult to integrate spatial data with other data holdings. This is not to say that there have not been developments in specialist GIS software such as the inclusion of powerful geostatistical methods and of heuristics for solving real-world problems such as the location–allocation problem. However at the same time there has been a movement of spatial data and GIS methods into mainstream data processing. This leads to the next theme of the book which is that greater attention is now being paid to methods for dealing with 2D data by the computer science community and by the IT industry in general. The early work on GIS was undertaken by a small group of academics, many with a background in geography. This work laid the foundations for modern GIS software but produced systems which were designed to handle small datasets. As soon as the problems which are being handled become very large, then small inefficiencies start to matter and you need systems which are designed for handling large datasets. One of the big changes in GIS in recent years has been much greater use of methods and data structures developed for handling large, non-spatial databases efficiently. This has sometimes involved adapting the techniques for 2D data, and sometimes converting the data to a 1D form so that the techniques can be applied. The third theme, and the factor which is probably driving most change in current GIS, is the explosion in the use of spatial data thanks to the Internet, satellite navigation, and mobile computing. It is difficult to underestimate the importance of all these technologies. When the first edition of this book was published in 2002, GIS was still something which was largely undertaken by trained staff using a desktop computer. The software was becoming easier to use and more widely available but it was still not something designed for the casual user. Now, in 2013, route planning, browsing maps and finding information about places are all tasks which are undertaken everyday by members of the public using what are effectively specialised GIS systems. The user interfaces are simple and intuitive, not least because these systems only offer a limited set of functionality. However, the functionality is powerful and useful and hence is widely used. These systems also draw on databases which are global in coverage but local in scale and also have to serve large numbers of concurrent users often accessing the system over a wireless network connection.

I originally considered concluding this book with a section on Future Directions in GIS but on reflection decided this would be futile. Change is too rapid and too unpredictable and anything I write now would simply cause amusement when read in 10 years' time. One is for certain though. GIS has become part of everyday life and is here to stay and I look forward to updating this book in another 10 years and seeing just what has changed.

Glossary

This glossary contains a set of informal definitions of some of the terms introduced in this book: Common GIS terms, such as vector and raster, which are described in standard GIS textbooks are not included.

1-cell: One of the names for the lines connecting nodes in the **link and node** data structure for representing vector data. Others are **chain**, **edge**, **link**, **segment** and **arc**.

Address: Number which uniquely identifies one unit of storage in computer memory. Computer circuitry is designed so that information can be moved between the CPU and all addresses with equal speed.

Algorithm: A procedure or set of steps to provide a solution to a problem. It should be possible to prove that the algorithm will always provide a solution. See **heuristic**.

Application domain model: A representation of some element or elements of the real world in a form which is comprehensible to experts in a particular application area. A map may be regarded as an application domain model.

Arc: One of the names for the lines connecting nodes in the **link and node** data structure for representing vector data. Others are **chain**, **edge**, **1-cell**, **segment** and **link**.

Array: A general purpose computer data structure which can store a list of data elements. Individual elements can be accessed extremely efficiently in memory.

Assembly language: Low-level computing language which translates directly into machine code. Assembly languages are therefore specific to a particular CPU or class of CPUs.

Balanced tree: A tree data structure in which the number of children below any node is kept as even as possible. This keeps the depth of the tree to a minimum and makes searches more efficient on average.

Big O notation: Informally, if an algorithm has $O(n^2)$ complexity, then as the number of objects to be handled (n) increases, the number of operations to be performed (and hence the approximate time taken) increases as the square of n. More formally if $f(n) = O(g(n))$, $g(n)$ is always less than $f(n)$ times a constant value for values of n above a second constant value. So $n^2 + 4$ is $O(n^2)$ because $n^2 + 4$ is always less than $2n^2$ for values of n above 4.

Binary search tree: A data structure used for storing and searching sorted data. Each subtree has the property that all nodes to the left of the root are smaller than or equal to the root, and all nodes to the right are larger than or equal to the root. See **tree**.

Binary tree: Any tree data structure in which each node had a maximum of two children.

Bit: Binary digit (0 or 1 in base 2 arithmetic). The smallest unit of storage within a computer file or in memory.

Bit interleaving: A technique for combining the X and Y coordinates of a location into a single spatial key. The coordinates are represented as binary integers. A new value is created by taking 1 bit at a time from the X and Y coordinates in turn. See **Morton code**.

Bit shifting: Operation which moves the **bits** in a binary **word** or **byte** to the left or right. If the bits represent a binary integer, a shift of one bit is equivalent to multiplying or dividing by 2.

Branch: A **node** in a **tree** which has further nodes beneath it.

B-tree: A tree data structure in which each node can have a lot of child nodes. This makes it suitable for storage on the disk because each disk access can retrieve an entire node, and the number of accesses is minimised.

Byte: Unit of storage consisting of 8 bits.

Cache: An area of disk or memory used to store information in case it is needed again by a program. Caching operates at all levels from the saving of information from Internet searches on the disk to the storage of sets of machine code and data within caches in the **CPU**.

Centroid: A point which is used to label an area. It must lie within the area, and should ideally be located near the centre.

Chain: One of the names for the lines connecting nodes in the **link and node** data structure for representing vector data. Others are **link, edge, 1-cell, segment** and **arc**.

Clustering: Grouping data values on disk storage in such a way that data items which are likely to be needed at the same time are stored close to each other on the disk and can therefore be retrieved in a small number of disk accesses.

Complexity: A measure of the efficiency of an algorithm. It is expressed as a function which describes how the time or storage needed to solve a problem increases as the size of the problem increases. Most commonly described using **big-O notation**, which describes the worst-case behaviour of the algorithm.

Computability: A term which describes whether a problem can be solved in an effective manner. It can be applied to abstract mathematical problems, but is most commonly applied to whether it is possible to solve the problem using a computer. Computability is often studied using a **Turing machine** as the model of a computer because this avoids issues related to the design of individual computers.

Computer precision: The number of significant digits which can be reliably stored and used in calculations in a computer **word**.

Conceptual computational model: A representation of an **application domain model** in a form which is potentially representable on a computer, but which is not specific to any particular language or package. The vector and raster models may both be regarded as conceptual domain models.

CPU: Central processing unit of a computer. Capable of performing operations on data held in the memory of the computer. As well as standard arithmetic operations, such as addition, usually able to perform operations such as bit shifting. Often contains a small number of **registers**.

Data model: A representation of a selection of objects or phenomena from the real world in a form which can potentially be stored on a computer. The data model itself contains no details of the computer implementation and is purely a conceptual construct. In non-spatial databases, the relational model is commonly used. The vector and raster models are commonly used for spatial data.

Data structure: Describes any structure used to organise data in the memory of a computer. Common structures include the **array** and various forms of **tree**. Specialised structures have been developed for the storage of spatial data.

DIME: Dual independent map encoding. A **data structure** for the storage of address-based street information developed by the US Bureau of the census.

Disk block: A collection of sectors on a disk. Data are always retrieved from the disk in whole blocks and so the block size determines the minimum amount of data which can be retrieved in a single disk access.

Disk sector: Data on disk drives is laid out in a series of concentric circles called tracks. Each track is subdivided into a series of sectors.

Double precision: The storage of floating point numbers in a 8 byte (64 bit) **word**. This means that calculations only have a **precision** of approximately 15 significant digits.

Edge: (1) In graph theory, the connection between nodes in a **graph**. (2) In GIS one of the names for the lines connecting nodes in the **link and node** data structure for representing vector data. Others are chain, link, 1-cell, segment and arc.

Face: In graph theory, the area defined by the edges of a **graph**. If a graph has no cycles (closed loops) there is only 1 face.

Floating point: Name for the method used to store numbers with a fractional part in the computer. The digits are stored in the **mantissa**, while the position of the decimal point is stored in the **exponent**.

Geometrical data: Data which store information about the location of vector objects. Sometimes loosely referred to as geometry. See **Topological data**.

Georelational: A name applied to GIS systems in which the locational (or **geo**graphical) data are stored separately from the attribute data, which is held in a relational database.

Graph: A representation of the set of connections between items. The items are represented as points, called **nodes**, and the connections between them as lines, called **edges**. The areas which are defined by the network of edges are called **faces**. A map of a road network can be regarded as a graph. **Tree** data structures can also be represented as graphs.

Heuristic: A procedure or set of steps which will generally produce a solution to a problem. In contrast to an **algorithm**, a heuristic does not have to solve the entire problem, and does not guarantee to provide good solutions in all cases.

Hexadecimal: Storage of numbers to base 16. Letters A–F are used to represent the digits above 9.

Interpolation: The estimation of an unknown value of a variable or parameter from known values. Strictly, the known values must lie on either side of the unknown value. If the unknown value exceeds the known ones, or lies outside the spatial extent of the known ones, the process is known as extrapolation.

k-d tree: A data structure used for point data. The area is subdivided according to the X coordinate of points at even levels in the tree, and the Y coordinate of points at odd levels. The structure allows for efficient spatial searches of points. See **tree**.

Leaf: A **node** in a **tree** which does not have any further nodes beneath it.

Line segment: Straight line connecting two points, forming part of the digitised representation of a line.

Link: One of the names for the lines connecting nodes in the **link and node** data structure for representing vector data. Others are chain, edge, 1-cell, segment and arc.

Linked list: A data structure for storing lists of items. Each member of the list contains a **pointer** to the next. In a doubly linked list, members also contain pointers to the previous member of the list.

Logarithm: The logarithm of a number n, is the value to which a second number (the base of the logarithm) must be raised to give a value of n. If $n = 2^m$ then $m = \log_2(n)$ that is, m is the logarithm of n to base 2. A logarithmic algorithm is one which is O(log n) in **big-O notation**.

Logical computational model: A representation of a **conceptual computational model** in a form which is specific to one language or package, but not to any particular make of computer. The data structure used by a particular GIS for vector data may be regarded as a conceptual computational model and the term is synonymous with the use of the term data structure in this book and in the GIS literature.

Lossless compression: Any method of reducing the size of a file without losing any of the information in it.

Lossy compression: Any method of reducing the size of a file which discards some information in order to achieve greater compression. Good lossy compression methods are based on retaining the most important information.

Machine code: The data which is used to tell the CPU what needs to be done. Made up of **opcodes**, which describe the operations to be carried out, and information on where the data to be processed on can be found.

Mantissa: The actual digits of a **floating point** number.

Minimum bounding rectangle: The smallest rectangle which completely encloses a spatial object.

Minimum enclosing rectangle: Another name for minimum bounding rectangle.

Monotonic section: Part of a line in which the X or Y coordinates always either increase or decrease.

Morton code: A spatial key for an object formed by alternately taking bits from its X and Y coordinates. When applied to pixels, the key forms the basis of addressing used in **quadtrees**. More generally provides an efficient way of indexing objects by their location.

Node: (1) In GIS, a point joining two lines in the **link and node data structure** for vector data. (2) In graph theory, an item in a **graph** which is connected to other nodes via **edges**. Also referred to as a **vertex**. Tree data structures are often represented using **graphs**, in which case both **leaves** and **branches** are represented by graph nodes.

Non-leaf node: Any node in a tree which has one or more child nodes. An alternative name for a branch node.

Opcode: The name for the individual instructions which tell the CPU what actions to perform, such as the addition of two numbers or performing a bit-shift operation.

Paging: A term used to describe the movement of data between file storage and memory in units of a fixed size called pages. There is a complex relationship between the size of the page, which is a function of the operating system, and the size of the **disk block**, which is a function of the disk drive architecture.

Planar enforcement: A topological rule which considers all lines in a vector layer to lie in the same plane. This means that lines must intersect at a **node** if they cross. One consequence of this is that if the lines define areas, any point must lie in one of the areas, and can only lie in one area.

Plane sweep algorithm: An algorithm which processes objects in X or Y coordinate order. If objects are processed in decreasing Y order, the action of the algorithm is often visualised as a horizontal line sweeping down the map.

Pointer: Information which connects data in one part of a **data structure** with data elsewhere in the structure.

Polygon: (1) Mathematically, a plane figure with three or more straight sides. (2) In GIS, often used as a synonym for area features, since these are represented as polygons in vector GIS.

Polynomial: A function in which at least one of the terms is raised to a power. For example, $y = ax^2 + bx + c$.

Procedure: In programming, the name for a self-contained section of a program. For example, sqrt(n) might be a procedure, which is passed the value of n as an argument, and returns its square root as a result.

Pseudocode: A description of an algorithm or heuristic in sufficient detail that a programmer could implement it in any suitable language. This may take the form of a written description in English, but more commonly uses a syntax based on a simplified version of a programming language.

Quadratic: (1) A **polynomial** function in which at least one term is raised to the power of 2, and no terms are raised to a higher power. (2) A quadratic algorithm is one which is O(n^2) in **big-O notation**.

Quadtree: A data structure which stores spatial data successively dividing the area of interest into quarters until each element is internally uniform. Commonly applied to raster data, in which the subdivision continues until all pixels within an area have the same value. Can provide compression of raster data and is important as an efficient spatial index.

Recursion: The name given to the style of programming in which a procedure calls itself. Many **algorithms** in which a problem is broken down into successively smaller parts, lend themselves naturally to recursion.

Register: Part of CPU which allows data which is used repeatedly to be stored in the CPU instead of being read from memory each time.

Relation: The proper name for a table in a relational database.

Root: The starting node of a **tree data structure**. Since trees are normally drawn with their **edges** hanging down, the root is usually at the top of the diagram.

R-tree: A tree data structure used to store the definition of rectangular areas for the purpose of forming an index to spatial data.

Run length encoding: A data structure for the storage of raster data. Instead of storing the value of every pixel, runs of pixels with the same value are identified. The pixel value and length of each run are then stored.

Secondary storage: General term for the storage of data on a permanent medium, such as CD or magnetic disk. Data must be read into memory from secondary storage before it can be used.

Segment: (1) The name used for the lines connecting nodes in the **DIME** data structure for representing vector data. (2) The straight line connecting two points as part of a link.

Single precision: The storage of floating point numbers in a 4 **byte** (32 bit) **word**. This means that calculations only have a **precision** of approximately 6–7 significant digits.

Sliver polygon: Small **polygons** formed when two almost identical lines are overlaid. This can arise if the same line is digitised twice by mistake, since the two versions will be similar but will differ slightly in the location of the digitised points.

Space-filling curve: Name given to a class of mathematical curves which cover the 2D plane, and which can be recursively defined at successively finer spatial scales. Since there is no limit to the recursion, at the limit the curves fill the space and become effectively 2D.

Toggle: In programming, a variable which flips between two possible values.

Topological data: Data which stores information about the connections between vector objects. Sometimes loosely referred to as topology. See **Geometrical data**.

Tree: A data structure consisting of **nodes** or **vertices**, connected by **edges**. The structure has a single **node** which is the **root**. This will normally have two or more child nodes, which in turn may have child nodes of their own. Trees represent data in a hierarchical manner, and the number of levels in the tree is normally O(log n). Queries which only have to visit each level, rather than each node, will therefore often have O(log n) **complexity**.

Tuple: The proper name for one row in a relational table. The word is also used for other collections of related data items, such as a row in an array and can be preceded by the number of items, for example, a 3-tuple.

Turing machine: A conceptual computing machine proposed by Alan Turing consisting of a read/write head fed by an infinite length of tape. The tape contains symbols from a fixed set, with each symbol causing the machine to move the tape to the left or right and/or write a symbol to the tape. Devised as a thought experiment to help in thinking about issues of computability, Turing machines continue to be useful in theoretical computer science.

Vertex: (1) In GIS, any of the points along a digitised line except the **nodes** at each end. (2) In graph theory, an alternative term for **node**.

Word: Unit of storage consisting of several bytes. 2 and 4 byte words are commonly used to store integers, and words of 4 or more bytes are used to store **floating point** numbers.

References

Abel, D.J. and Mark, D.M. 1990. A comparative analysis of some two-dimensional orderings. *International Journal of Geographical Information Systems* 4(1), 21–31.

Arge, L. 1997. External-memory algorithms with applications in GIS. In van Kreveld, M., Nievergelt, J., Roos, T. and Widmayer, P. (eds), *Algorithmic Foundations of Geographic Information Systems*. Lecture Notes in Computer Science 1340. Berlin: Springer-Verlag, 213–254.

Arostegui, Jr. M.A., Kadipasaoglu, S.N. and Khumawala, B.M. 2006. An empirical comparison of Tabu search, simulated annealing, and genetic algorithms for facilities location problems. *International Journal of Production Economics* 103(2), 742–754.

Baars, M. 2003. *A Comparison between ESRI Geodatabase Topology and Laser-scan Radius Topology*. TU Delft Internal Report. Available from http://repository.tudelft.nl/ (accessed 6/11/2012).

Bentley, J.L. and Ottmann, T.A. 1979. Algorithms for reporting and counting geometric intersections. *IEEE Transactions on Computers* C-28,9, 643–647.

Berry, J.K. 1993. *Beyond Mapping: Concepts, Algorithms and Issues in GIS*. Fort Collins, CO: GIS World Books.

Berry, J.K. 1995. *Spatial Reasoning for Effective GIS*. Fort Collins, CO: GIS World Books.

Blakemore, M. 1984. Generalization and error in spatial databases. *Cartographica* 21, 131–139.

Blum, C. and Roli, A. 2003. Metaheuristics in combinatorial optimization: Overview and conceptual comparison. *ACM Computing Surveys* 35(3), 268–308.

Bowyer, A. and Woodwark, J. 1983. *A Programmer's Geometry*. London: Butterworth.

Bugnion, E., Roos, T., Wattenhofer, R. and Widmayer, P. 1997. Space filling curves and random walks. In van Kreveld, M., Nievergelt, J., Roos, T. and Widmayer, P. (eds), *Algorithmic Foundations of Geographic Information Systems*. Lecture Notes in Computer Science 1340. Berlin: Springer-Verlag, 199–211.

Burrough, P.A. and McDonnell, R.A. 1998. *Principles of Geographical Information Systems*. Oxford: Oxford University Press.

Chen, H.-L. and Chang, Y.-I. 2005. Neighbour-finding on space-filling curves. *Information Systems* 30, 205–226.

Chen, Q. 2007. Airborne lidar data processing and information extraction. *Photogrammetric Engineering and Remote Sensing* 73(2), 109–112.

Comer, D. 1979. The ubiquitous B-tree. *ACM Computing Surveys* 11(2), 123–137.

Cooper, L. 1972. The transportation-location problem. *Operations Research* 20(1), 94–108.

Cormen, T.H., Leiserson, C.E. and Rivest, R.L. 1990. *Introduction to Algorithms*. Cambridge, MA: MIT Press.

Dale, P. 2005. *Introduction to Mathematical Techniques Used in GIS*. Boca Raton, FL: CRC Press.

de Berg, M., van Kreveld, M., Overmars, M. and Schwarzkopf, O. 1997. *Computational Geometry*. Berlin: Springer.

DeCandia, G., Hastorun, D., Jampani, M., Kakulapati, G., Lakshman, A., Pilchin, A., Sivasubramanian, S., Vosshall, P. and Vogels, W. 2007. Dynamo: Amazon's highly available key-value store. *Proceedings of SOSP'07*, October 14–17, 2007, Stevenson, Washington, USA. pp. 205–220. http://www.read.seas.harvard.edu/~kohler/class/cs239-w08/decandia07dynamo.pdf (accessed 10 June 2013).

DeMers, M.N. 1999. *Fundamentals of Geographic Information Systems* (2nd edition). Chichester: Wiley.

De Simone, M. 1986. Automatic structuring and feature recognition for large scale digital mapping. *Proceedings Auto Carto London*, London: AutoCarto London Ltd, 86–95.

Devereux, B. and Mayo, T. 1992. Task oriented tools for catographic data capture. *Proceedings AGI'92*, p.2.14.1-2.14.7, London: Westrade Fairs Ltd.

Douglas, D.H. 1974. It makes me so cross. Reprinted in Peuquet, D.J. and Marble, D.F. (eds). 1990. *Introductory Readings in Geographical Information Systems*. London: Taylor & Francis.

Dutton, G. 1999. *A Hierarchical Coordinate System for Geoprocessing and Cartography*. Lecture Notes in Earth Science 79. Berlin: Springer-Verlag.

Dyer, C.R. 1982. The space efficiency of quadtrees. *Computer Graphics and Image Processing* 19, 335–348.

ESRI. 1998. *ESRI Shapefile Technical Description*. http://www.esri.com/library/white-papers/pdfs/shapefile.pdf (accessed 1 Mar 2013).

Evans, D. 2011. *Introduction to Computing*. http://www.computingbook.org/ (accessed 1 Mar 2013).

Farr, T.G., Rosen, P.A., Caro, E. et al. 2007. The shuttle radar topography mission. *Reviews of Geophysics* 45(2), RG2004.

Fisher, P.E. and Tate, N.J. 2006. Causes and consequences of error in digital elevation models. *Progress in Physical Geography* 30(4), 467–489.

Fisher, P.F. 1993. Algorithm and implementation uncertainty in viewshed analysis. *International Journal of Geographical Information Systems* 7(4), 331–347.

Fisher, P.F. 1997. The pixel—A snare and a delusion. *International Journal of Remote Sensing* 18(3), 679–685.

Foley, J.D., van Dam, A., Feiner, S.K. and Hughes, J.F. 1990. *Computer Graphics: Principles and Practice*. Reading, MA: Addison-Wesley.

Franke, R. 1982. Scattered data interpolation: Tests of some methods. *Mathematics of Computation* 38(157), 181–200.

Fu, L., Sun, D. and Rilett, L.R. 2006. Heuristic shortest path algorithms for transportation applications: State of the art. *Computers and Operations Research* 33(11), 3324–3343.

Gahegan, M.N. 1989. An efficient use of quadtrees in a geographical information system. *International Journal of Geographical Information Systems* 3(3), 201–214.

Gallay, M., Lloyd, C.D., McKinley, J. and Barry, L. 2013. Assessing modern ground survey methods and airborne laser scanning for digital terrain modelling: A case study from the Lake District, England. *Computers and Geosciences* 51, 216–227.

Gardiner, V. 1982. Stream networks and digital cartography. *Cartographica* 19(2), 38–44.

Geisberger, R., Sanders, P., Schultes, D. and Vetter, C. 2012. Exact routing in large road networks using contraction hierarchies. *Transportation Science* 46(3), 388–404.

Gold, C.M. 1992. Surface interpolation as a Voronoi spatial adjacency problem. *Proceedings Canadian Conference on GIS*; Ottawa, ON, 1992, pp. 419–431. (available on http://www.voronoi.com, accessed 6 Mar 2013).

Goodchild, M.F. and Grandfield, A.W. 1983. Optimizing raster storage: An examination of four alternatives. *Proceedings Auto Carto 6*, Ottawa, 1, 400–407.

Goodchild, M.F. and Kemp, K.K. (eds). 1990. *NCGIA Core Curriculum in GIS*. Santa Barbara, CA: National Center for Geographic Information and Analysis, University of California.

Goodchild, M.F. and Lam, N. 1980. Areal interpolation: A variant of the traditional spatial problem. *Geoprocessing* 1, 297–312.

Greene, D. and Yao, F.F. 1986. Finite-resolution computational geometry. *Proceedings 27th IEEE Symposium on Foundations of Computer Science*, pp. 143–152, Toronto.

Haggett, P. and Chorley, R.J. 1969. *Network Analysis in Geography*. London: Edward Arnold.

Harel, D. 2000. *Computers Ltd. What They Really Can't Do*. Oxford: Oxford University Press.

Hart, P.E., Nilsson, N.J. and Raphael, B. 1968. A formal basis for the heuristic determination of minimum cost paths. *IEEE Transactions on Systems Science and Cybernetics SSC4* 4(2), 100–107.

Healey, R.G. 1991. Database management systems. In Maguire, D.J., Goodchild, M.F. and Rhind, D.W. *Geographical Information Systems—Principles and Applications*. Harlow: Longman. Vol. 1, 251–267.

Hogg, J., McCormack, J.E., Roberts, S.A., Gahegan, M.N. and Hoyle, B.S. 1993. Automated derivation of stream channel networks and related catchment characteristics from digital elevation models. In Mather P.M. (ed.), *Geographical Information handling—Research and Applications*. New York: John Wiley, 207–235.

Huang, C.-W. and Shih, T.-Y. 1997. On the complexity of the point in polygon algorithm. *Computers and Geosciences* 23(11), 109–118.

Hutchinson, M.F. 1989. A new procedure for gridding elevation and stream line data with automatic removal of spurious pits. *Journal of Hydrology* 106(3–4), 211–232.

Lawder, J.K. and King, P.J.H. 2001. Querying multi-dimensional data indexed using the Hilbert space-filling curve. *ACM Sigmod Record* 30(1), 19–24.

Jenson, S.K. and Domingue, J.O. 1988. Extracting topographic structure from digital elevation model data for geographic information system analysis. *Photogrammetric Engineering and Remote Sensing* 54, 1593–1600.

Jones, C.B. 1997. *Geographical Information Systems and Computer Cartography*. Harlow: Longman.

Keating, T., Phillips, W. and Ingran, K. 1987. An integrated topologic database design for geographic information systems. *Photogrammetric Engineering and Remote Sensing* 53, 1399–1402.

Knuth, D. 1998. *Sorting and Searching* (Vol. 3 *The Art of Computer Programming*) (2nd edition). Reading, MA, Addison-Wesley.

Kumler, M.P. 1994. An intensive comparison of triangulated irregular networks (TINs) and digital elevation models (DEMs). *Cartographica* 31(2), 1–99.

Lam, N. 1983. Spatial interpolation methods: A review. *The American Cartographer* 10(2), 129–149.

Leigh, C.L., Kidner, D.B. and Thomas, M.C. 2009. The use of LiDAR in digital surface modelling: Issues and errors. *Transactions in GIS* 13(4), 345–361.

Li, R., Guo, R., Xu, Z. and Fen, W. 2012. A prefetching model based on access popularity for geospatial data in a cluster-based caching system. *International Journal of Geographical Information Science* 26(10), 1831–1844.

Liu, C.-M., Kao, R.-L. and Wang, A.-H. 1994. Solving location–allocation problems with rectilinear distances by simulated annealing. *Journal of the Operational Research Society* 45(11), 1304–1315.

Liu, X. 2008. Airborne LiDAR for DEM generation: Some critical issues. *Progress in Physical Geography* 32(1), 31–49.

Lloyd, C. and Atkinson, P.M. 2006. Deriving ground surface digital elevation models from LiDAR data with geostatistics. *International Journal of Geographical Information Science* 20(5), 535–563.

Louwsma, J., Tijssen, T. and van Oosterom, P. 2003. Topology under the microscope. *Geo Connexion* 2003, 29–30.

Mark, D.M. 1975. Computer analysis of topography: A comparison of terrain storage methods. *Geografisker Annaler* 57A, 179–188.

Mark, D.M. 1979. Phenomenon-based data-structuring and digital terrain modelling. *Geo-Processing* 1, 27–36.

Mark, D.M. 1984. Automated detection of drainage networks from digital elevation models. *Cartographica* 21, 168–178.

Mark, D.M. 1989. Neighbour-based properties of some orderings of two-dimensional space. *Geographical Analysis* 22(2), 145–157.

Marks, D., Dozier, J. and Frew, J. 1984. Automated basin delineation from digital elevation data. *Geo-Processing* 2, 299–311.

Mather, P.M. 1999. *Computer Processing of Remotely-Sensed Images: An Introduction* (2nd edition). Chichester: Wiley.

Mizukoshi, H. and Aniya, M. 2002. Use of contour-based DEMs for deriving and mapping topographic attributes. *Photogrammetric Engineering and Remote Sensing* 68(1), 83–93.

Moon, B., Jagadish, H.V., Faloutsos, C. and Saltz, J.H. 2001. Analysis of the clustering properties of the Hilbert curve. *IEEE Transaction on Knowledge and Data Engineering* 13(1), 124–141.

Moore, I.D., O'Loughlin, E.M. and Burch, G.J. 1988. A contour-based topographic model for hydrological and ecological applications. *Earth Surface Processes and Landforms* 13, 305–320.

MSDN. 2012. *Spatial Data Types Overview*. http://msdn.microsoft.com/en-us/library/bb964711.aspx (accessed 6/11/2012).

Nagy, G. 1980. What is a "good" data structure for 2-D points? In Freeman H. and Pieroni G.G. *Map Data Processing*. New York: Academic Press, 119–136.

NCGIA. 2000. *The NCGIA Core Curriculum in GIScience*. http://www.ncgia.ucsb.edu/giscc/ (accessed 6 Mar 2013).

Nievergelt, J. 1997. Introduction to geometric computing: From algorithms to software. In van Kreveld, M., Nievergelt, J., Roos, T. and Widmayer, P. (eds), *Algorithmic Foundations of Geographic Information Systems*. Lecture Notes in Computer Science 1340. Berlin: Springer-Verlag, 1–19.

Nievergelt, J. and Widmayer, P. 1997. Spatial data structures: Concepts and design choices. In van Kreveld, M., Nievergelt, J., Roos, T. and Widmayer, P. (eds), *Algorithmic Foundations of Geographic Information Systems*. Lecture Notes in Computer Science 1340. Berlin: Springer-Verlag, 153–197.

OGC. 2006. *OpenGIS Implementation Specification for Geographic Information—Simple Feature Access—Part 1: Common Architecture.* OGC 06-103r3. http://www.opengeospatial.org/standards/sfa (accessed 6 Mar 2013).

Park, S.C. and Shin, H. 2002. Polygonal chain intersection. *Computers and Graphics* 26, 341–350.

Peucker, T.K., and Chrisman, N. 1975. Cartographic data structures. *American Cartographer* 2(1), 55–69.

Peucker, T.K., Fowler, R.J., Little, J.J. and Mark, D.M. 1978. The triangulated irregular network. *Proceedings American Society of Photogrammetry Digital Terrain Models Symposium*, St. Louis, MO, 516–540.

Peuquet, D.J. 1981a. An examination of techniques for reformatting digital cartographic data, Part I, The raster-to-vector process, *Cartographica* 18(1), 34–48.

Peuquet, D.J. 1981b. An examination of techniques for reformatting digital cartographic data, Part II, The vector-to-raster process, *Cartographica* 18(3), 21–33.

Peuquet, D.J. 1984. A conceptual framework and comparison of spatial data models. *Cartographica* 21(4), 66–113.

Peuquet, D.J. and Marble, D.F. 1990. *Introductory Readings in Geographical Information Systems.* London: Taylor & Francis.

Phibbs, C.S. and Luft, H.S. 1995. Correlation of travel time of roads versus straight line distance. *Medical Care Research and Review* 52(4), 532–542.

Rosenfeld, A. 1980. Tree structures for region representation. In Freeman H. and Pieroni G.G. *Map Data Processing.* New York: Academic Press.

Rosenfeld, A. and Kak, A. 1982. *Digital Picture Processing.* London: Academic Press.

Saalfeld, Alan. 1987. It doesn't make me nearly as CROSS. *International Journal of Geographical Information Systems* 1(4), 379–386.

Samet, H. 1990a. *Applications of Spatial Data Structures: Computer Graphics, Image Processing, and GIS.* Reading, MA: Addison-Wesley.

Samet, H. 1990b. *The Design and Analysis of Spatial Data Structures.* Reading, MA: Addison-Wesley.

Samet, H. and Aref, W.G. 1995. Spatial data models and query processing. In Kim, E. (ed.), *Modern Database Systems: The Object Model, Interoperability and Beyond.* New York: Addison-Wesley/ACM Press, 338–360.

Schirra, S. 1997. Precision and robustness in geometric computations. In van Kreveld, M., Nievergelt, J., Roos, T. and Widmayer, P. (eds), *Algorithmic Foundations of Geographic Information Systems.* Lecture Notes in Computer Science 1340. Berlin: Springer-Verlag, 255–287.

Shekhar, S. and Chawla, S. 2003. *Spatial Databases: A Tour.* Upper Saddle River, NJ: Prentice-Hall.

Sibson, R. 1981. A brief description of natural neighbour interpolation. In Barnett, V. (ed.), *Interpreting Multivariate Data*, Chichester: Wiley, 21–36.

Skidmore, A.K. 1989. A comparison of techniques for calculating gradient and aspect from a gridded digital elevation model. *International Journal of Geographical Information Systems* 3(4), 323–334.

Smith, T.R., Menon, S., Star, J.L. and Estes, J.E. 1987. Requirements and principles for the implementation and construction of a large-scale geographic information system. *International Journal of Geographical Information Systems* 1(1), 13–31.

Stolze, K. 2003. SQL/MM spatial: The standard to manage spatial data in relational database systems. *Proceedings of BTW 2003: Database Systems for Business,*

Technology and Web. Leipzig, Germany. http://doesen0.informatik.uni-leipzig. de/proceedings/paper/68.pdf (accessed 6/11/2012).

Tarboton, D.G. 1997. A new method for the determination of flow directions and upslope areas in grid digital elevation models. *Water Resources Research* 33(2), 309–319.

Taud, H., Parrot, J. F. and Alvarez, R. 1999. DEM generation by contour line dilation. *Computers & Geosciences* 25(7), 775–783.

Teng, T.A. 1986. Polygon overlay processing: A comparison of pure geometric manipulation and topological overlay processing. *Proceedings 2nd International Symposium on Spatial Data Handling*. Williamsville, NY: International Geographical Union, 102–119.

Theobald, D.M. 2001. Topology revisited: Representing spatial relations. *International Journal of Geographical Information Science* 15(8), 689–705.

Theobald, D.M. and Goodchild, M.F. 1990. Artifacts of TIN-based surface flow modelling. *Proceedings GIS/LIS '90, Volume 2.* Bethesda, MD: ASPRS/ACSM, 955–967.

Toutin, T. 2008. ASTER DEMs for geomatic and geoscientific applications: A review. *International Journal of Remote Sensing* 29(7), 1855–1875.

US Bureau of the Census. 1990. Technical description of the DIME system. In Peuquet and Marble, *Introductory Readings in GIS*, London: Taylor & Francis, 100–111.

van Kreveld, M. 1997. Digital elevation models and TIN algorithms. In van Kreveld, M., Nievergelt, J., Roos, T. and Widmayer, P. (eds), *Algorithmic Foundations of Geographic Information Systems*. Lecture Notes in Computer Science 1340. Berlin: Springer-Verlag, 37–78.

van Kreveld, M., Nievergelt, J., Roos, T. and Widmayer, P. (eds). 1997. *Algorithmic Foundations of Geographic Information Systems*. Lecture Notes in Computer Science 1340. Berlin: Springer-Verlag.

van Oosterom, P. 1999. Spatial access methods. In Longley, P.A., Goodchild, M.F., Maguire, D.J. and Rhind, D.W. (eds), *Geographical Information Systems 1: Principles and Technical Issues*. New York: Wiley and Sons, 385–400.

Wang, F. and Sun, Y. 2002. Spatial object clustering and buffering. *IEEE Multimedia* 2012, 26–42.

Waugh, T.C. 1986. A response to recent papers and articles on the use of quadtrees for geographic information systems. *Proceedings 2nd International Symposium on Spatial Data Handling*. Williamsville, NY: International Geographical Union, 33–37.

White, D. 1978. A design for polygon overlay. In Dutton G. (ed.), *First International Advanced Study Symposium on Topological Data Structures for Geographic Information Systems, Volume 6.* Cambridge, MA: Laboratory for Computer Graphics and Spatial Analysis.

White, M. 1984. Tribulations of automated cartography and how mathematics helps. *Cartographica* 21, 148–159.

Wilson, J.P. and Gallant, J.C. (eds). 2000. *Terrain Analysis—Principles and Applications.* Chichester: John Wiley and Sons.

Wise, S.M. 1988. Using contents addressable filestore for rapid access to a large cartographic database. *International Journal of Geographical Information Systems* 2(2), 11–120.

Wise, S.M. 1995. Scanning thematic maps for input to geographic information systems. *Computers and Geosciences* 21(1), 7–29.

Wise, S.M. 1997. The effect of GIS interpolation errors on the use of DEMs in geo-morphology. In Lane, S.N., Richards, K.S. and Chandler, J.H. (eds), *Landform Monitoring, Modelling and Analysis*, Chichester: Wiley.

Wise, S.M. 1999. Extracting raster GIS data from scanned thematic maps. *Transactions in GIS* 3(3), 221–237.

Wise, S.M. 2000a. Data modelling in GIS—Lessons from the analysis of digital eleva-tion models (Guest editorial). *International Journal of Geographical Information Science* 14(4), 313–318.

Wise, S.M. 2000b. Assessing the quality for hydrological applications of digital eleva-tion models derived from contours. *Hydrological Processes* 14(11–12), 1909–1929.

Worboys, M.F. and Duckham, M. 2004. *GIS: A Computing Perspective* (2nd edition). Boca Raton, FL: CRC Press.

Xie, Q., Zhang, Z. and Ravada, S. 2012. In-database raster analytics: Map algebra and parallel processing in oracle spatial georaster. In Shortis, M. and Madden, M. (eds), *Proceedings XXII ISPRS Congress, Melbourne, Australia (International Archives of the Photogrammetry, Remote Sensing and Spatial Information Sciences, Volume XXXIX-B4)*. Göttingen: Copernicus Publication, 91–96.

Young, R. 2009. *How Computers Work*. http://www.fastchip.net/howcomputerswork/p1.html (accessed 1 Mar 2013).

Yu, S., van Kreveld, M. and Snoeyink, J. 1996. Drainage queries in TINs: From local to global and back again. In Kraak, M.-J. and Molenaar, M. (eds), *Advances in GIS II: Proceedings 7th Int. Symposium on Spatial Data Handling*. London: Taylor & Francis, 829–855.

Index

8-bit byte, 15, 120
    contents of, 24
    storage of 2 in, 198
16-bit computer, 14

A

a-machine, 12
A* algorithm, 226, 234; *see also* Dijkstra
      algorithm
    for best-first search, 232
    characteristics, 230, 231
    heuristic properties, 231
    minimal effect on, 230
    weighted, 233
Active segment list (ASL), 90, 91, 92
Algorithm, 3
    convex polygons, 186
    coordinate system, 186
    design, 5
    elevation, 188
    graph of equation, 187
    gridded DEM, 195–199
    hydrological analysis,
        192–195
    intercept, 187
    interpolated estimates, 190
    interpolation process, 191
    polynomial equations, 189
    slope, 189
    straight line equation, 186
    for surfaces, 185
    third-order polynomial, 191
    TIN, 185
Algorithm complexity, 109
    measurement of, 116
    value of functions, 112
Algorithm efficiency, *see* Efficiency
      of algorithms
Alternating transportation-location
      heuristic (ATL heuristic), 280
    greedy approach, 282
    for location-allocation problem, 294
    steps in, 281

Annealing, 284
ARC/INFO, 3
    storage of area data in, 63
ARC software, 36, 62, 63
Areas, 10, 32, 69. Multiple polygons;
      Single polygons
    administrative, 47
    attributes for, 46
    locational data for, 51
    multiple, 46
    sliver polygons, 47
    topographic map and features, 40
    woodland and non-woodland, 48
Array, 10, 17, 123
    element identification in, 124, 125
    $O(n^2)$ storage efficiency, 127, 128
    raster, 125, 142
    storage in computer memory, 124,
      126, 142
    storing lists of information, 43
    uses, 124
ASL, *see* Active segment list (ASL)
Assembly languages
    using C language, 25
    instructions, 25
    for Intel x86 processors, 24, 25
    raster analysis, 122, 123
    translated into machine code, 26
ASTER GDEM, 166, 167, 168
ATL heuristic, *see* Alternating
      transportation-location
      heuristic (ATL heuristic)
AX register, 14

B

B-tree, 249, 261, 262
Balanced binary tree, 257
    leaf nodes, 258
    range search, 259, 260
Ballistic range, 12, 13
Bentley–Ottmann algorithm, 89, 92
    line intersection algorithm, 89
    plane sweep algorithm, 91